SCIENCE REFERENCE SOURCES

SCIENCE REFERENCE SOURCES

FRANCES BRIGGS JENKINS

Fifth Edition

The M.I.T. Press
Cambridge, Massachusetts, and London, England

MIT Press

0262600021

JENKINS
SCIENCE REFERENCE

Second printing, December 1970
Third printing, April 1973

ISBN 0 262 10003 7 (hardcover)

ISBN 0 262 60002 1 (paperback)

Library of Congress catalog card number: 73-95001

Printed in the United States of America

PREFACE

This selected list of science reference sources is a complete
revision of a compilation used in the courses in science reference
service at the University of Illinois Graduate School of Library
Science. In the seventeen years since the first preliminary edition was prepared, our attention has been directed to the use of
the publication in courses in bibliography at other colleges and
universities, in orientation programs, etc. as well as by the staff
and users of library collections. We are gratified that it has
proven helpful to those librarians and science literature users
who are attempting to keep up with the wealth of science reference
materials currently available. It is hoped that this latest edition
will prove of equal or greater value than its predecessors.

A work of this type does not pretend to be comprehensive.
As in former editions, the titles listed are those one might ex-
pect to find in the general reference collection of a large research
library. Nor has an attempt been made to provide a list of "Best
Books"; in some instances citations have been included merely
because they represent a type of useful reference aid. This bibliography presents titles selected from the variety of science
reference materials available on June 1, 1969. Since a knowledge
of general reference books has been assumed, such works have not
been included, although they may very well provide valuable as-
sistance in science reference work.

Arrangement of the titles in this bibliographical manual is in
sections by subject. Presentation of general science reference
works is followed by highly selected lists of general aids in mathe-
natics, physics, chemistry, astronomy, earth sciences, biological sciences, medical sciences, agricultural sciences and engineering sciences. Within each section the entries are usually
arranged according to type of reference book, e.g. guides to the
literature, bibliographies, indexes and abstracts, reviews and
surveys, histories, dictionaries and encyclopedias, handbooks,
directories, special tools, etc. Each entry has been given an
item number to facilitate discussion and use. The call number
assigned by the University of Illinois Library is also listed.

An effort has been made to economize on space whenever pos-
sible without detracting from the clarity of the presentation, by

using standard abbreviations and omitting place of publication when the publisher is well known. In this edition, reference to the inclusion of the title in Winchell's Guide to Reference Books, eighth edition, or its first supplement, is indicated by the designator "Wi-" followed by the Winchell code number; reference to citation in Walford's Guide to Reference Materials, second edition, Volume 1: Science and Technology is shown by the abbreviation "Wa" and the page number. In many instances, reference to a comment on the title in other sources is included, using the abbreviations listed after the Preface.

Grateful appreciation is expressed to the many librarians and library school students who furnished facts and made suggestions for this manual. The corrections and suggestions which were gratefully received from the users of previous editions have been given careful consideration. The reference and departmental librarians at the University of Illinois Library have been particularly helpful with comments on new publications. Library School students who deserve words of appreciation include the graduate assistants who helped to verify entries, check call numbers, and locate copies of the works. Of this group my very special gratitude goes to Mary Elvin Eckert for her sustained interest and efforts through the dreary task of rechecking bibliographic citations, proof reading, and concern with problems which attend the compilation of such a work. The author reserves the right to take credit for all of the errors in the compilation. Mrs. Allen B. Wilson is thanked for the final typing of the manuscript.

Again, continued interest in making the list of maximum value is encouraged, and we welcome your future suggestions.

Frances Briggs Jenkins

Champaign, Illinois
June 1, 1969

ABBREVIATIONS

Listed below are the chief abbreviations used in the bibliographic citations. This list does not include the generally accepted abbreviations.

Am. Doc.: American Documentation
Am. Lib. Assn.: American Library Association
Am. Chem. Soc.: American Chemical Society
Am. Scientist: American Scientist
Bibl. Doc. Term.: Bibliography, Documentation, Terminology
Biblio. Soc. of Am.: Bibliographical Society of America
Booklist: The Booklist section of the Booklist and Subscription
    Books Bulletin
Chem. & Eng. News: Chemical & Engineering News
Choice: Choice; Books for College Libraries
Clearinghouse: Clearinghouse for Federal Scientific and Technical
    Information, Department of Commerce, Washington
Coll. & Res. Lib.: College & Research Libraries
Geol. Soc. Am.: Geological Society of America
G.P.O.: Government Printing Office
J. Chem. Doc.: Journal of Chemical Documentation
J. Chem. Ed.: Journal of Chemical Education
Jour. of Doc.: Journal of Documentation
Lib. J.: Library Journal
Lib. Q.: Library Quarterly
L.C. Info. Bull.: Library of Congress Information Bulletin
LRTS: Library Resources and Technical Services
Math. of Comp.: Mathematics of Computation
Math. Rev.: Mathematical Reviews
Mech. Eng.: Mechanical Engineering
Med. Lib. Assn. Bull.: Bulletin of the Medical Library Association
New Tech. Bks.: New York Public Library. New Technical Books.
Rec. Math. Mag.: Recreational Mathematics Magazine
S-H Book News: Stechert-Hafner Book News
Sat. Rev.: Saturday Review
Sci. Am.: Scientific American
Sci. Info. Notes: Scientific Information Notes
Sci. Ref. Notes: Scientific Reference Notes
Sp. Libs.: Special Libraries
Sp. Lib. Assn.: Special Library Association
Sub. Bks. Bull.: The Subscription Books Bulletin section of The
    Booklist and Subscription Books Bulletin

Tech. Bk. Rev. Index: Technical Books Review Index
Univ. Pr.: University Press
U.S.G.S.: United States Geological Survey
Wilson Lib. Bull.: Wilson Library Bulletin
Wi-___: Winchell, C.M. Guide to Reference Books. 8th ed.
   (followed by code number for title)
Wi-1___: First Supplement, 1965-66 to above title
Wa p.___: Walford, A.J. Guide to Reference Material. 2d ed. v.1:
   Science and Technology (followed by page number)

CONTENTS

SECTION A. GENERAL SCIENCE

Science has been defined by James B. Conant in his On Understanding Science (Yale University Press, 1947, p. 98) as "that portion of accumulative knowledge in which new concepts are continuously developing from experiment and observation and lead to further experimentation and observation".

GUIDES TO THE LITERATURE

A1. American Scientific Books. Bowker, 1962-65. (Continued in **American Book Publishing Record. Annual Cumulative.** 1965-   .)
016.5 Am36
  See: Lib. J. 87:3426-7, 1962; Choice 1:542-3, 1 965. Wi-EA20.

A2. Dissertation Abstracts. Section B: The Sciences and Engineering.
University Microfilms, v.27, 1966-   .
  See: Wi-1AI2.

A3. Downs, R. B. & Jenkins, F. B., eds. Bibliography: Current State
and Future Trends. Univ. of Ill. Pr., 1967. (Reprint of Library
Trends Jan. & Apr., 1967 with index) 016.016 D75b
  See: Lib. J. 93:1118-9, 1968; Choice 4:1363, 1968.

A4. Hawkins, R. R. Scientific, Medical and Technical Books Published in the U.S.A. 2d ed., Bowker, 1958. 016.5 H31s
  See: Sub. Bks. Bull. 55:397, 1959. Wi-EA11. Wa p. 14.

A5. Maichel, K. Guide to Russian Reference Books. Vol. V: Science,
Technology, and Medicine. Hoover Institution, Stanford Univ., 1967.
016.9147 M24g
  See: Lib. J. 93:2846, 1968; Med. Lib. Assn. Bull. 56:347-8, 1968.

A6. Malclès, L. N. Les Sources du Travail Bibliographique. Tome 3:
Bibliographies spécialisées (Sciences Exactes et Techniques). Geneva,
Droz, 1958. 010 M24s
  See: Jour. of Doc. 15:156-9, 1959. Wi-AA301. Wa p. 6.

A7. Malinowsky, H. R. Science and Engineering Reference Sources;
a Guide for Students and Librarians. Libraries Unlimited, 1967.
016.5 M295s
  See: Lib. J 93:55, 1968; Sp. Libs. 59:208, 1968.

1

A8. Schutze, G. Bibliography of Guides to the S-T-M Literature: Scientific, Technical, Medical. N.Y., 1958- . Supplements, 1962- . 016.5 Sch8b
　　See: Sp. Libs. 59:117, 1968.

A9. Walford, A.J., ed. Guide to Reference Material. 2d ed. Vol 1: Science and Technology. London, Library Assn., 1966. 025.52 W174g
　　See: Lib. J. 91:3687, 1966. Wi-1EA2.

A10. Winchell, C.M. Guide to Reference Books. 8th ed., Am. Lib. Assn., 1967. Supplements, 1968- . 025.51 M944n
　　Supplements, compiled by E. P. Sheehy, appear in January and July issues of College & Research Libraries.
　　See: Coll. & Res. Libs. 28:290, 1967; Lib. J. 93:3106, 1968.

See also such publications as:
　　Bennett, M. Science and Technology: A Purchase Guide for Branch and Small Public Libraries. Pittsburgh, Carnegie Library, 1963. Supplements, 1964- . 016.5 B43s
　　Bonn, G.S. Science-Technology Literature Resources in Canada. Ottawa, National Research Council, 1966. 507.2 B64s
　　Houghton, B. Technical Information Sources. A Guide to Patents, Standards and Technical Reports Literature. Archon, 1967. 600 H81t
　　Jackson, I.H. ed. Acquisition of Special Materials. Palo Alto, Altoan Pr., 1967. 025.2 C153a

BIBLIOGRAPHIES, INDEXES, ABSTRACTS

Guides to bibliographic aids

A11. Besterman, T. A World Bibliography of Bibliographies. Lausanne, Switzerland. Societas Bibliographica. 1965-66. 5v. 011 B465
　　See: Wi-1AA3.

A12. International Federation for Documentation. Abstracting Services in Science, Technology, Medicine, Agriculture, Social Sciences, Humanities. The Hague, 1965. q016.0294 In8ab
　　Supplemented by listings in FID News Bulletin (010.8205 FI)

A13. U.S. Library of Congress. A guide to the World's Abstracting and Indexing Services in Science and Technology. 1963. 016.5 Un36gu
　　See: Sci. Ref. Notes 10:3, (Jan./Apr.) 1963. Wi-EA62.

A14. Reuss, J.D. Repertorium commentationium... Göttingen. Dieterich, 1801-1821. 16v. (v. 1, Historia naturalis, generalis et zoologia; v. 2, Botanica et mineralogia; v. 3, Chemia et res metallica; v. 4, Physica; v. 5, Astronomia; v. 6, Oeconomia, v. 7, Mathesis, mechanica, hydrostatica, hydraulica, hydrotechnia, aerostatica, pneumatica technologia, architectura civilia, scientia navalis, scientia militaris; v. 8, Historia; v. 9, Philologia, linguae, scriptores graeci, scriptores latini, litterae elegantiores, poesis, rhetorica, ars antiqua, pictura, musica; v. 10-16, Scientia et ars medica et chirurgica) Reprinted by B. Franklin, 1961. A500 R31r
  See: Wi-EA15. Wa p. 7.

A15. Royal Society of London. Catalogue of Scientific Papers, 1800-1900. Cambridge, 1867-1925. 19 v. Subject Index. Cambridge, 1908-1914. 3 v. in 4. (v. 1-6 indexes literature of 1800-63; v. 7-8 indexes literature 1864-73; v. 9-11 indexes literature 1874-83; v. 12, supplements previous volumes; v. 13-19 indexes literature 1884-1900.) (Subject index: v. 1, Pure mathematics; v. 2, Mechanics; v. 3:1, Physics: Generalities, heat, light, sound; v. 3:2, Physics: Electricity and magnetism). 016.5 R81
  Reprinted by Scarecrow Pr. 1968.
  See: Wi-EA16. Wa p. 8.

A16. International Catalogue of Scientific Literature, 1st-14th Annual Issues. London, Harrison, 1902-19. 254 v. (Each annual issue in 17 v.: A, Mathematics; B, Mechanics; C, Physics; D, Chemistry; E, Astronomy; F, Meterology, terrestrial magnetism; G, Mineralogy, petrology, crystallography; H, Geology; J, Geography; K, Palaeontology; L, General biology; M, Botany; N, Zoology; O, Human anatomy; P, Physical anthropology, Q, Physiology, experimental psychology, pharmacology, experimental pathology; R, Bacteriology.) Reprinted by Johnson, 1965-66. 016.5 In8
  See: Lib. J. 93:4091, 1968. Wi-EA12. Wa p. 6.

A17. U.S. National Library of Medicine. Index-catalogue of the Library... 1st ser. v. 1-16, 1880-95; 2d ser. v. 1-21, 1896-1916; 3d ser. v. 1-10, 1918-32; 4th ser. v. 1-11, 1936-55; 5th ser. v. 1-3, 1959-61. (Originally: Index-Catalogue of the Library of the Surgeon General's Office. Continued by: U.S. National Library of Medicine. Catalog. v. 1, 1948-   . and Index Medicus. v.1, 1960-   .) q016.61 Un3i
  See: Wi-EJ13. Wa p. 180.

A18. France. Centre National de la Recherche Scientifique. Bulletin Signalétique. v. 1, 1940- . (Sections: 1, Mathématiques; 2, Astronomie. Astrophysique. Physique du Globe; 3, Physique I; 4, Physique II; 5, Physique nucléare; 6, Structure de la matière; 7, Chimie 1; 8, Chimie II; 9, Sciences de l'ingénieur; 10, Sciences de la terre I; 11, Sciences de la terre II; 12, Biophysique. Biochimie; 13, Sciences pharmacologiques. Toxicologie; 14, Microbiologie. Virus. Bacteriophages. Immunologie. Génétique; 15, Pathologie générale et experimentale; 16, Biologie et physiologie animales; 17, Biologie et physiologie vegetales; 18, Sciences agricoles. Zootechnie. Phytiatrie et Phytopharmacie. Aliments et industries alimentaires; 19, Philosophie. Sciences Humaines; 20, Psychologie. Pédagogie; 21, Sociologie; 22, Histoire des sciences et des techniques; 23, Esthetique. Archeologie; 24, Sciences du language) Title 1940-1956 was Bulletin Analytique, which had absorbed Bibliographie Mensuelle de l'Astronomie in 1948. A500 F84b
  See: Wi-EA71. Wa p. 4.

A19. Referativnyi Zhurnal. Moscow. Akad. Nauk SSSR. 1953- .
Titles, coverage, frequency and indexing of series vary. Sections of some series are issued as separates. Other individual titles are not part of a series. (Catalogued as separates)
  See: Maichel, K. Guide to Russian Reference Books. v. 5 p. 360
Wi-EA72. Wa p. 7.

A20. Science Citation Index. Institute for Scientific Information, 1963- . Issued in three quarterly and an annual cumulation; available weekly on magnetic tape. Permuterm Subject Index, 1967- . (Annual) q016.505 Sc2
  See: Library Trends 16:374-86, 1968; LRTS 9:478-82, 1965 and 12:415-34, 1968; Chem. & Eng. News 42:55-6, (Aug. 31) 1964; Jour. of Doc. 21:139-41, 1965; Lib. J. 89:2735-7, 1964; Science 145:142-3, 1964. Wa p. 8.

A21. Pandex. N. Y., CCM Information Sciences, 1967- . Microfiche; Quarterly with annual cumulations. Available on magnetic tape.
  Pandex Current Index of Scientific and Technical Literature. 1969-
A biweekly printed version of Pandex
  See: Am. Doc. 19:357-8, 1968; Coll. & Res. Libs. 29:72, 1968;
Sp. Libs. 58:728-30, 1967; Sci-Tech News 23:19, 1969.

A22. Australian Science Index. Melbourne, Commonwealth Scientific and Industrial Research Organization, v. 1, 1957- . 505 AUST
  See: Wa p. 16.

A23. Current Contents: Physical Sciences. Institute for Scientific Information, v. 1, 1961- . . Title 1961-67: Current Contents: Your Weekly Guide to Space and Physical Sciences. 505 CURC
See: Wa p. 203.

A24. John Crerar Library. Catalog. G.K. Hall, 1967. Author-title catalog, 35 v. Classified catalog, 42v. Subject index to classified catalog, 1v.

A25. Masters Theses in Pure and Applied Sciences. Purdue Univ., 1955/56- . (title varies) 016.62 P972m
See: Wi-EA33.

See also such compilations as:
    Downs & Jenkins, Bibliography: Current State and Future Trends. (No. A3)
    "Science Abstracting Services" issue of Library Trends, January, 1968.

HISTORIES

A26. Sarton, G. Horus: a Guide to the History of Science. Chronica Botanica, 1952. 509 Sa7g
See: Coll. & Res. Libs. 13:399-400, 1952. Wi-EA176. Wa p. 44.

A27. Kronick, D.A. A History of Scientific and Technical Periodicals; the Origins and Development of the Scientific and Technological Press, 1665-1790. Scarecrow, 1962. 070.486 K92h
See: Coll. & Res. Libs. 24:174, 1963; LRTS 7:122-4, 1963. Wa p. 29 (note).

A28. Sarton, G. A History of Science. Harvard Univ. Pr., 1952-59. 2v. 509 Sa7hi
See: Booklist 49:120, 1952. Wa p. 42.

A29. Sarton, G. Introduction to the History of Science. Williams & Wilkins, 1927-48. 3v. in 5. (Bibliographic supplement in Isis) 509 Sa7i
See: Science 128:641-4, 1958. Wi-EA181. Wa p. 42.

A30. Singer, C.J. A Short History of Scientific Ideas to 1900. Oxford, 1962. (Based on his A Short History of Science published in 1941) 509 Si6sc
See: Science 131:405, 1960. Wa p. 43.

5

A31. Source Books in the History of Science series. Harvard Univ.
Pr., 1929-   . (Early volumes published originally by McGraw-Hill)
   See: Science 115:3, (Mar 21) 1952.

A32. Taton, R. ed. History of Science. Basic Books, 1963-6. 4v.
English translation of his Histoire générale des Sciences. Presses
Universitaires. 1957-64. 3v. in 4) 509 T18aEp
   See: French edition reviewed in Science 127:973, 1958; 130:382+, 1959;
137:418, 1962. English edition reviewed in Lib. J. 89:123, 1964.
Wa p. 43.

A33. Thorndike, L. A History of Magic and Experimental Science.
Columbia Univ. Pr., 1923-58. 8v. (First 6 volumes published by
Macmillan) 509 T39h
   See: Science 129:90-1, 1959. Wi-EH44. Wa p. 43.

A34. Thornton, J.L. & Tully, R.I.J. Scientific Books, Libraries, and
Collectors. 2d ed., London, Library Assn., 1962. 016.5 T39s
   See: Science 139:897, 1963. Wi-EA177. Wa p. 9.

A35. History of Science: an Annual Review of Literature, Research and
Teaching. Cambridge, Heffer, v. 1, 1962-   509 H628
   See: Wa p. 42.

A36. Isis; an International Review Devoted to the History of Science
and Its Cultural Influences. Johns Hopkins Univ., v. 1, 1913-
505 ISI

   See also such surveys as:
      Holland, H.E. Academic Libraries and the History of Science.
      Univ. of Ill. Graduate School of Library Science. Occasional
      Papers. No. 91, Jan. 1968.
      Neu, J. The History of Science. In Downs & Jenkins, eds.
      Bibliography: Current State and Future Trends. p. 438-454.
      (No. A3)

DICTIONARIES & ENCYCLOPEDIAS

Guides

A37. Marton, T.W. Foreign-language and English Dictionaries in the
Physical Sciences and Engineering; a Selected Bibliography, 1952-63.
Wash., G.P.O., 1964. (U.S. National Bureau of Standards Misc. Pub.
258) 016.503 M36f
   See: Coll. & Res. Libs. 26:56, 1965. Wi-EA94. Wa p. 18.

A38. Turnbull, W.R. Scientific and Technical Dictionaries; an Annotated Bibliography. San Bernardino, Calif., Bibliothek, 1966-
v.1: Physical Science and Engineering. 1966. Supplement, 1968.
016.503 T84s
  See: RQ 7:48, 1967; New Tech. Bks. 52:255, 1967; Lib. J. 93:2982, 1968.

A39. UNESCO. Bibliography of Interlingual Scientific and Technical Dictionaries. 4th ed. Paris, 1961. Supplement, 1965. (Supplemented by listings in Bibliography, Documentation, Terminology) 016.503 H73b
  See New Tech. Bks. 47:115, 1962. Wi-EA96. Wa p. 18.

A40. Walford, A.J. ed. A Guide to Foreign Language Grammars and Dictionaries. London, Library Association, 1964. 016.4 W14g
  See: W1-AE107.

General works

A41. Ackner, J. Pocket Encyclopedia of Physical Sciences: Astronomy, Chemistry, Geology, Meteorology, Physics. Golden Pr., 1968.
  See: Booklist 65:619, 1969.

A42. The Book of Popular Science. Grolier Society, 1964. 10v.
500 B644
  See: Sub. Bks. Bull. 55:166, 1958.

A43. Chambers' Technical Dictionary. 3d ed. rev. with supplement. Macmillan, 1958. 603 C355
  See: Sub. Bks. Bull. 55:261, 1959. Wi-EA99. Wa p. 21.

A44. Compton's Illustrated Science Dictionary. Rev. ed., Chicago, 1969. 503 C738d
  See: Sub. Bks. Bull. 60:272+, 1963. Wi-EA104. (Review of 1st ed.)

A45. Graham, E.C., ed. The Basic Dictionary of Science. Edited in Basic English for the Orthological Institute. Macmillan, 1966.
503 G76b

A46. Handwörterbuch der Naturwissenschaften. Jena, Fischer, 1931-35. 10v. and indexes. 503 H192
  See: Wi-EA84. Wa p. 18 (note).

A47. Harper Encyclopedia of Science. Rev. 2d ed. Harper & Row, 1967. 1 or 2 v. 503 H234
  See: Choice 4:1215, 968; Lib. J. 93:741, 1968; RQ 7:90, 1967.

A48. McGraw-Hill Encyclopedia of Science and Technology. 2d ed.
McGraw-Hill, 1965. 15v. Updated by McGraw-Hill Yearbook of
Science and Technology. 1962- . 503 M178
  Supplemented by McGraw-Hill Basic Bibliography of Science and
Technology. 1966. 016.5 M178
    See: Sub. Bks. Bull. 63:1158-1161, 1967 and 64:793-802, 1968;
Lib. J. 92:2146, 1967; Sci. Am. 214:138-140, (June) 1966. Wi-1EA11.

A49. Van Nostrand's Scientific Encyclopedia. 4th ed. Van Nostrand,
1968. 503 V338
    See: Choice 5:1417, 1969; Coll. & Res. Libs. 30:84, 1969.

Yearbooks

A50. McGraw-Hill Yearbook of Science and Technology. McGraw-Hill,
1962- . 503 M1781
    See: Lib. J. 92:2146, 1967.

A51. Britannica Yearbook of Science and the Future. Encyclopedia
Britannica, 1968- .

A52. Encyclopedia Science Supplement. Grolier, 1965-   (An annual
supplement to the Grolier Encyclopedia)
    See: Science 152:917, 1966; Lib. J. 92:1577, 1967.

A53. Science Year: The World Book Science Annual. Field Enterprises,
1965- . 031 W8942s
    See: Sub. Bks. Bull. 62:725-7, 1966; Science 152:917, 1966. Wi-1EA12.

Special works

A54. Acronyms and Initialisms Dictionary. 2d ed. Gale, 1965. 421.8 Ac7

A55. Ballentyne, D.W.G. & Walker, L.E.Q. Dictionary of Named Effects and Laws in Chemistry, Physics, and Mathematics. 2d ed.
Macmillan, 1961. 503 B21d
    See: Wi-EA79. Wa p. 21.

A56. Clark, G.L., ed. The Encyclopedia of Microscopy. Reinhold,
1961. 578.03 C54e
    See: Science 133:1916, 1961; Sub. Bks. Bull. 58:541-2+, 1962.
Wi-ED22. Wa p. 81.

A57. Clark, G. L. The Encyclopedia of Spectroscopy. Reinhold, 1960. 535.84 **En19**

A58. De Sola, **R.** Abbreviations Dictionary. Rev. ed., Meredith, 1967.
  See: <u>Lib. J.</u> **93:535-6**, 1968.

A59. Hough, **J.N.** Scientific Terminology. Rinehart, 1953. 503 H81s
  See: <u>New Tech. Bks.</u> 38:27, 1953. Wi-EA101

A60. Ruffner, F.G. Jr. & Thomas, R.C. eds. Code Names Dictionary: A Guide to Code Names, Slang, Nicknames, Journalese and Similar Terms: Aviation, Rockets and Missiles, Military, Aerospace, Meteorology, Atomic Energy, Communications and others. Gale, 1963. 423 R83c
  See: Wi-CI204. Wa p.258.

<u>Foreign language</u>

  French

A61. Cusset, F. English-French and French-English Technical Dictionary. 2d ed. rev. Chemical Pub., 1957. 603 C96e
  See: Wi-EA106. Wa p. 24.

A62. DeVries, L. French-English Science Dictionary ... 3d ed. McGraw-Hill, 1962. 503 D49f
  See: Wi-EA107. Wa p.24.

  German

A63. De Vries, L. German-English Science Dictionary   3d ed. McGraw-Hill, 1959. 503 D49G
  See: Wi-EA108. Wa p. 22.

A64. Dorian, A.F., comp. Dictionary of Science and Technology, English-German. American Elsevier, 1967. 503 D73d
  See: <u>Lib. J.</u> 93:1125, 1968.

A65. Leibiger, O.W. & Leibiger, I.S. German-English and English-German Dictionary for Scientists. Edwards, 1950. 503 L53g
  See: Wi-EA112. Wa p. 23.

A66. Webel, A. German-English Dictionary for Technical, Scientific and General Terms   3d ed. Dutton, 1953. 503 W38g
  See: Wi-EA115. Wa p.23.

Hungarian

A67. Angol-magyar Műszaki Szótár. English-Hungarian Technical Dictionary. Heinman, 1959. q603 A277
  See: Wi-EA118. Wa p. 28.

A68. Magyar-angol Műszaki Szótár. Hungarian-English Technical Dictionary. Heinman, 1957. q603 M277.
  See: EA117. Wa p. 28.

Italian

A69. Denti, R. Dizionario Technico; Italiano-Inglese, Inglese-Italiano. 6th ed., Milan, Hoepli, 1965. 603 D43d
  See: Wi-EA119. Wa p. 25.

A70. Gatto, S. Dizionario Technico Scientifico Illustrato: Italiano-Inglese, Inglese-Italiano. Milan, Ceschina, 1960.
  See: Wi-EA120.

Russian

A71. Blum, A. Concise Russian-English Scientific Dictionary for Students and Research Workers. Pergamon, 1965. 503 B62c

A72. Bray, A. Russian-English Scientific and Technical Dictionary. International Univ. Pr., 1945. 603 B73r
  See: Wi-EA127. Wa p. 25 (note).

A73. U.S. Library of Congress. Soviet Russian Scientific and Technical Terms, A Selected List. G. P.O., 1963. 503 Un32s
  See: Wa p. 27.

A74. Zalucki, H. Dictionary of Russian Technical and Scientific Abbreviations: with Their Full Meaning in Russian, English and German. Elsevier, 1968.
  See: Lib. J. 93: 4129, 1968; Choice 6:198, 1969.

A75. Zimmerman, M.G. Russian-English Translators Dictionary, a Guide to Scientific and Technical Usage. Plenum, 1967. 503 T78r
   See: New Tech. Bks. 52:201, 1967.

   Spanish

A76. Castilla's Spanish and English Technical Dictionary. Philosophical, 1958. 2v. 603 C278
   See: Wi-EA132. Wa p. 25.

HANDBOOKS & TABLES

A77. Brady, G.S. Materials Handbook. 9th ed. McGraw-Hill, 1963. 603 B72m
   See: Wi-CH129. Wa p. 207.

A78. Elsevier's Lexicon of International and National Units. Elsevier, 1964. 389.103 E172
   See: Wi-CH191.

A79. Handbook of Chemistry and Physics. Cleveland, Chemical Rubber Pub., 1st ed., 1913- . (annual) 541.9 H191
   See: Wi-ED46. Wa p. 95.

A80. Kaye, G.W.C., comp. Tables of Physical and Chemical Constants, and Some Mathematical Functions. 13th ed., Wiley, 1966. (Earlier editions compiled by Kaye, G.W.C. and Laby, T.H., now prepared under the direction of an editorial committee.) 530.8 K18t
   See: New Tech. Bks. 52:72, 1967.

A81. Landolt, H.H. Landolt-Börnstein Zahlenwerte und Funktionen aus Physik, Chemie, Astronomie, Geophysik und Technik. 6th ed., Berlin, Springer. 1950- (in progress) 541.9 L23p

A82. Landolt, H.H. Landolt-Börnstein Zahlenwerte und Funktionen aus Naturwissenschaften und Technik. Neue Serie. Berlin, Springer, 1961- . (in progress)
   See: National Standard Reference Data System. News June 1967.
Wi-EA138. Wi-1EA22. Wa p. 41.

A83. National Research Council. International Critical Tables. McGraw-Hill, 1926-1933. 7v. and index. 502 N21i
   See: Wi-EA139. Wa p. 41 (note).

A84.   Palmer, E. L.   Fieldbook of Natural History.   McGraw-Hill, 1949.
570 P18f
   See: Science 110:129, 1949. Wa p. 52.

A85.   Smithsonian Institution. Smithsonian Physical Tables.   9th rev.
ed., 1954. (Smithsonian Misc. Colls., v. 120) 530.8 Sm6p
   See: Wi-EG40.

A86.   Zimmerman, O.T. & Lavine, I. Industrial Research Service's
Conversion Factors and Tables. 3d ed. Dover, N.H., 1961. 530 Z6i
   See: Wi-EG42. Wa p. 41.

   Note sections in:
      World Almanac and Book of Facts, 1868-
      Information Please Almanac, 1947-
      Statistical Abstracts of the U.S., 1879-

BIOGRAPHIES

A87.   American Men of Science. 11th ed. Bowker, 1965-67. Supplements,
1966-   . 925 Am31
   See: Lib. J. 92:2751, 1967. Wi-1EA34 and 1EA35. Wa p. 46.

A87a. A Biographical Dictionary of Scientists. Wiley, 1969.

A88.   Dictionary of Scientific Biography. Scribner. In preparation,
4-5 volumes projected, first two volumes planned for fall of 1969.
   See announcement in Physics Today 18:86, 1965.

A89.   Directory of British Scientists. London, Benn. 1963-
(annual) 925 D62
   See: Wi-EA186. Wa p. 45.

A90.   Ireland, N.O. Index to Scientists of the World, from Ancient to
Modern Times: Biographies and Portraits. Faxon, 1962. 016.925
Ir2i
   See: Sub. Bks. Bull. 59:835-6, 1963. Wi-EA194. Wa p. 45.

A91.   McGraw-Hill Modern Men of Science. McGraw-Hill, 1966. 925
M178
   See: Science 153:731, 1966; Subs. Bks. Bull. 64:793-802, 1968.
   Wi-1EA36.

A92. National Academy of Sciences. Biographical Memoirs. Wash.,
1877-   . (Publication program varies) 925 N21.
 See: Wa p. 47.

A93. Nobel Lectures; Including Presentation Speeches and Laureates
Biographies. Elsevier, 1964-   . (Series: Physics. 3v.; Chemistry.
3v.; Physiology-Medicine. 3v.; Literature. 1v.; Peace. 1v.)
 See: New Tech. Bks. 49:269, 1964. Wi-ED80.

A94. Poggendorff, J.C. Biographisch-literarisches Handwörterbuch
zur Geschichte der exacten Wissenschaften. Berlin, Verlag Chemie,
1863-1940. 6v. in 10. v.7, 1956-   . (Title varies slightly) 925 P 75
 v. 1-2 cover period to 1857; v. 3, 1858-83; v. 4, 1883-1904;
v. 5, 1904-22; v.6, 1923-31; v. 7, 1932-53.
 See: Wi-EA191. Wa p. 45.

A95. Royal Society of London. Biographical Memoirs of Fellows of
the Royal Society. v.1, 1955-   . (Continues Obituary Notices of
Fellows and "Obituary Notices" formerly included in its Proceedings)
925 R81b

A96. Turkevich, J. & Turkevich, L.B. comp. Prominent Scientists
of Continental Europe. Elsevier, 1968.
 See: Lib. J. 94:1130, 1969.

A97. World Who's Who in Science; a Biographical Dictionary of Notable
Scientists from Antiquity to the Present. Marquis, 1968.
 See: Coll. & Res. Libs. 30:78, 1969.

 See also:
     Biography Index; a Quarterly Index to Biographical Material in
        Books and Magazines. Wilson, v.1, 1946-   (Wi-AJ2)

DIRECTORIES

Guides

A98. International Federation for Documentation. Directories of
Science Information Sources, International Bibliography. The Hague,
1967. 016.06 In8b
 See: Sci. Info. Notes 9:22, (Oct.) 1967.

A99. U.S. Library of Congress. Directories in Science and Technology; a Provisional Checklist. G.P.O., 1964. 016.5058 Un3d
 See: Wi-EA149. Wa p. 35.

See also:
    Guide to American Directories. Prentice-Hall, v.1, 1954-

Information resources

A100. Aslib Directory: A Guide to Sources of Information in Great
Britain and Ireland. London, Aslib, 1928- . 027.23 A849a

A101. Directory of Computerized Information in Science & Technology.
Science Associates/International, 1968. (Looseleaf with supplements)
q010.78 D628
    See: Med. Lib. Assn. Bull. 57:87-8, 1969.

A102. Directory of Department of Defense Information Analysis Centers.
Wash., G.P.O., 1966. 010.78 Un38d
    See: Bibl. Doc. Term. 8:172, 1968.

A103. Directory of Federally Supported Information Analysis Centers.
Clearinghouse, 1968.
    See: Sci. Info. Notes 10:23, (June/July) 1968.

A104. Directory of Special Libraries and Information Centers. Gale,
1961- . 027.9 D59
    See: Wi-AB33.

A105. Directory of Special Libraries in Israel. Tel Aviv, Center of
Sci. & Tech. Info., 1964.
    See: Sci. Info. Notes 6:13, (June/July) 1964.

A106. International Directory of Back Issue Vendors: Periodicals,
Newspapers and Documents. 2d ed., Special Libraries Assn., 1968.
    See: Sp. Libs. 59:689, 1968. 655.5058 In8

A107. National Research Council of Canada. National Technical Information Services: World Wide Directory. 2d ed. F.I.D., 1968.
607.2 N212n

A108. UNESCO. World Guide to Science Information and Documentation
Services. Paris, 1965 010 Un32w
    See: Sp. Libs. 57:68, 1966. Wi-1EA25.

A109. U.S. Library of Congress. International Scientific Organizations; A Guide to their Library, Documentation and Information Service.
G.P.O., 1963. 506 Un32i
    See: Sci. Ref. Notes 5:10, (July/Oct.) 1963. Wi-EA162. Wa p. 35.

A110. U.S. National Referral Center. A Directory of Information Resources in the United States: Federal Government. G.P.O., 1967. 001 Un3d
  See: Lib. J. 92:3978, 1967. Sci. Info. Notes 9:20, (Oct. /Nov.) 1967.

A111. U.S. National Referral Center. A Directory of Information Resources in the United States: Physical Sciences, Biological Sciences, Engineering. G.P.O., 1965. 500 Un31s
  See: Am. Doc. 17:47-8, 1966; Coll. & Res. Libs. 26:404-5, 1965. Wi-1EA26. Wa p. 39 (note).

  See also:
    Battelle Memorial Institute. Directories of Specialized Information
      Centers and Services. Sp. Libs. 59:97, 1968.

Research institutions

A112. Battelle Memorial Institute. Directory of Selected Scientific Institutions in the USSR. Merrill, 1963. 506 B32d
  See: Sp. Libs. 55:121-2, 1964. Wi-EA153. Wa p. 37.

A113. Directory of Research Institutions and Laboratories in Japan. Tokyo, Society for Promotion of Science, 1964. 068.52 J27d
  See: Wa p. 38.

A114. Directory of Selected Research Institutes in Eastern Europe. Columbia Univ. Pr., 1967. 507.2 L72d
  See: Sci-Tech News 22:78-9, 1968; New Tech. Bks. 52:126, 1967.

A115. European Research Index; a Guide to Scientific and Industrial Research in Western Europe. Guernsey, Hodgson Ltd., 1965. 2v. 507.2 Eu74
  See: Coll. & Res. Libs. 28:66, 1967.

A116. Industrial Research Laboratories of the United States. 12th ed., Bowker, 1965. (Earlier editions in the NAS-NRC Publication series) 607.2 In27
  See: Wi-EA166.

A117. International Directory of Research and Development Scientists. Institute for Scientific Information, 1968. q 925 In82
  See: Med. Lib. Assn. Bull. 57:94-5, 1969; Lib. J. 93:2790, 1968.

15

A118. Museums Directory of the United States and Canada. 2d ed.
Am. Assn. of Museums and Smithsonian Institution, 1965. 069.058 M97
   See: Science 149:964, 1965. Wi-EA170.

A119. Research Centers Directory. Gale. 1st ed., 1960-   (Up-dated
quarterly by New Research Centers, 1965-   .) (title varies)
378 D62
   See: L.C. Info. Bull. 24:315, 1965. Wi-EA147.

A120. UNESCO. World Directory of National Science Policy-making
Bodies. v.1: Europe and North America. N.Y., 1966-   . 506 W893
   See: Bibliography Documentation Terminology 7:26, 1967; Lib. J.
92:2146, 1967; RQ 7:50-1, 1967.

A121. Yugoslav Scientific Research Directory. Clearinghouse, 1964.
507.2 Y91
   See: Sci. Info. Notes 6:13, (June/July) 1964.

   Scientific societies

A122. Bates, R.S. Scientific Societies in the **United States.** 3d ed.
Mass. Inst. Tech. Pr., 1965. 506-273 B31s
   See: Sp. Libs. 57:263, 1966. Wi-EA152. Wa **p. 39.**

A123. Battress, F.A. World List of Abbreviations. **3d ed.** London,
Leonard Hill, 1966. 060 B98w
   See: Sp. Libs. 57:669, 1966. Wi-1AC1.

A124. National Academy of Sciences. The Eastern European Academies
of Sciences; a Directory. Wash., 1963. 068.49 N21e
   See: Wi-EA157.

A125. Scientific and Learned Societies of Great Britain; a Handbook
Compiled from Official Sources. London, Unwin, 1st ed., 1884-
(annual, 1884-1939; irregular, 1951-   . title varies) 016.062 YE
   See: Wi-EA158. Wa p. 36.

A126. Scientific and Technical Societies in Japan. Tokyo, Maruzen
Book Co., 1962.

A127. Scientific and Technical Societies of the United States. 8th ed. National Academy of Sciences, 1968. (NAS-NRC Pub. 1499) A506 N21h
See: Choice 4:88, 1968.

Note information in such general works as:
  Encyclopedia of Associations. 1957-
  The Foundation Directory. 1960- . (Supplemented by
    Foundation News, 1960- . bimonthly)
  World of Learning. 1947- .
  Yearbook of International Organizations. 1952- .

SCIENTIFIC MEETINGS

Calendar

A128. International Congress Calendar. Brussels, Union of International Associations. 1961- . (annual) (Supplements appear in International Associations, 1961- .) 060 In82
See: Wi-AC8.

A129. Scientific Meetings. N.Y., Sp. Lib. Assn., v.1, 1957-
506 SCIM
See: Wi-EA160.

A130. U.S. Surgeon-General's Office. Library. Congresses: Tentative Chronological and Bibliographical Reference List of National and International Meetings of Physicians, Scientists, and Experts. G.P.O., 1938. Supplement, 1939. (Reprint from the Library's Index-catalogue, 4th series, 3:1-288, 1938) A060 Un3c
See: Wi-EA164.

A131. World List of Future International Meetings. Library of Congress, 1959- . 060 W893
See: Wi-AC22. Wa p. 40.

A132. World Meetings, U.S. and Canada. Newton Centre, Mass., Technical Meetings Information Service. v.1, 1963- . (Title change with v.5, April 1966) 605 TM
See: Lib. J. 89:2779, 1964.

A133. World Meetings, Outside U.S.A. and Canada. Newton Centre, Mass., Technical Meetings Information Service, 1968- . 605 TMW

See also:
Announcements in official publications of societies.
Science (list in each issue with cumulation at irregular intervals)
Chemical & Engineering News ("Calendar of Meetings and Events" at
irregular intervals with supplements in intervening issues)
Cruzat, G.S. Keeping Up with Biomedical Meetings. RQ 7:12-20,
1967 or Med. Lib. Assn. Bull. 56:132-7, 1968.

Publications

A134. Bibliographical Current List of Papers, Reports and Proceedings
of International Meetings. Brussels, Union of International Associations.
1961- . 060.5 UNII

A135. Bibliography of Proceedings of International Meetings. Brussels,
Union of International Associations. 1963- . 016.06 Un34b

A136. Directory of Published Proceedings. White Plains, N.Y.,
InterDok Corp. 1965- . 060.5 DI

A137. International Congresses and Conferences, 1840-1937: a Union
List of their Publications Available in Libraries of the United States
and Canada. Wilson, 1938. qAO60 In84
See: Wi-CJ138.

A138. Yearbook of International Congress Proceedings. Stechert-
Hafner, 1968- . (2v. in preparation)
See: Stechert-Hafner Book News 23:8, (Fall) 1968.

TECHNICAL WRITING

Guides to the literature

A139. American Medical Writers Ass'n. Directory of Free-Lance
Writers, Editors & Researchers. Arlington, Va., 1966.

A140. Dopkowski, P.L., ed. Selected Bibliography of Indexing in
Science and Technology. American Univ., 1963. 016.0295 D72s

A141. Philler, T.A. et al. An Annotated Bibliography on Technical
Writing, Editing, Graphics, and Publishing 1950-1965. Society of
Technical Writers & Publishers, 1966. 016.808066 An 78

A142. Rutgers University. Graduate School of Library Science.
Bibliography of Research Relating to the Communication of Scientific
and Technical Information. Rutgers Univ. Pr., 1967.
  See: Lib. J. 92:3975, 1967.

A143. Shank, R. comp. Bibliography of Technical Writing, 1945-1957.
Society for Technical Writers and Editors, 1958. 016.62 Sh1b

A144. UNESCO. Bibliography of Publications Designed to Raise the
Standard of Scientific Literature. Paris, 1963. 016.808066 Un3b
  See: Sp. Libs. 54:319-320, 1964. Wa p. 3.

   General

A145. American Standards Association. American Standards for
Periodical Title Abbreviations, Z39.5-1963. N.Y., 1963. 025.173
A512a
  See: Sci. Info. Notes 6:17, (June/July) 1964. Wa p. 29 (note).

A146. Arnell, A. Standard Graphical Symbols: a Comprehensive
Guide for Use in Industry, Engineering, and Science. McGraw-Hill,
1963. 620.00148 Ar62a

A147. Ehrlich, E.H. & Murphy, D. The Art of Technical Writing: a
Manual for Scientists, Engineers, and Students. Crowell, 1964.
808.066 Eh8a
  See: New Tech. Bks. 49:199, 1964.

A148. Freedman, G. A Handbook for the Technical and Scientific
Secretary. Barnes & Noble, 1967. 502 F87h

A149. Guidry, N. & Frye, K.B. Graphic Communication in Science.
Wash., NEQ Publications Sales, 1968.
  See: Chem. & Eng. News 46 :47, (July 29) 1968; Sp. Libs. 59:750,
1968.

A150. Laird, E.S. Engineering Secretary's Complete Handbook.
2d ed., Prentice-Hall, 1967. 651.02 L14e
  See: New Tech. Bks. 53:23, 1968.

A151. Thomas, J.F. A Guide for Authors on Manuscripts, Proof and
Illustration. 2d ed., Thomas, 1968. (Original publication, 1949,
authored by P.E.L. Thomas)

A152. Turner, R.P. Technical Writer's & Editors Stylebook. Sams, 1964. 808.066 T85t
  See: Lib. J. 90:2836, 1965.

A153. Weisman, H.M. Technical Correspondence. Wiley, 1968. 808.6 W43t
  See: Mechanical Engineering 90:77, 1968.

TRANSLATIONS

A154. American Translators Association. ATA Professional Services Directory. N.Y., 1965- . 410.28 Am3
  See: Wi-1AA18.

A155. Himmelsbach, C.J. & Boyd, G. A Guide to Scientific and Technical Journals in Translation. S.L.A., 1968. 016.505 H577g
  See: Lib. J. 93:2790, 1968; Coll. & Res. Lib. 30:83-4, 1969.

A156. Index Translationum. UNESCO, v. 1-31, 1932-1940. New Series v.1, 1948- 010.5 IN
  See: L.C. Info. Bull. 24:255, 1965. Wi-AA91.

A157. Monthly Index of Russian Accessions. v.1, 1948- 015.47 Un32m
  See: Wi-AA673.

A158. Translations Register-Index. S.L.A. Translations Center. 1967- (semi-monthly) 605 TR
  See: Lib. J. 92:2505, 1967; Med. Lib. Assn. Bull. 55:534-5, 1967- (semi-monthly) 605 TR

A159. Translators and Translations: Services and Sources in Science and Technology. 2d ed. S.L.S., 1965. 410.28 T687
  See: Lib. J. 90:3260-1, 1965. Wi-1AA19.

A160. U.S. Clearinghouse for Federal Scientific and Technical Information. Technical Translations. v.1-18, 1959-1967. (Superseded Translation Monthly, v.1-4, 1955-58, which had absorbed Library of Congress' Bibliography of Translations from Russian Scientific and Technical Literature in 1957) U.S. government-sponsored technical translations are announced in U.S. Government Research & Development Reports beginning January 10, 1968. 016.6 Un32t
  See: Sp. Libs. 58:730, 1967.

A161. U.S. National Science Foundation. Providing U.S. Scientists with Soviet Science Information. Rev. ed., 1962. 016.5094 Un3p
  See: Wi-EA146.

A162. World Index of Scientific Translations. v.1, 1967-   (Quarterly) 016.5 W893
  See: Stechert-Hafner Book News 22:41, 1967.

INDUSTRIAL INFORMATION

A163. MacRae's Blue Book. v.1, 1881-   (Title varies) 670 M244
  See: Wi-EA143.

A164. Special Libraries Association. Sources of Commodity Prices. N.Y., 1960. 338.5 Sp3s
  See: Wi-CH146.

A165. Statistics Sources; a Subject Guide to Data on Industrial, Business, Social, Educational, Financial, and other Topics. 2d ed., Gale, 1966. 016.31 St2
  See: Wi-1CH28.

A166. Thomas' Register of American Manufacturers. Thomas Pub. Co., 1906- . (Annual) q670 T36
  See: Wi-CH332.

RESEARCH & DEVELOPMENT REPORTS

A167. Nuclear Science Abstracts   Atomic Energy Commission. v.1, 1948- . 539 N88
  See: Wi-EI202. Wa p. 217.

A168. Scientific and Technical Aerospace Reports (STAR). National Aeronautics and Space Administration. v.1, 1963- . 016.6294 Sc27
  See: Wi-EI20.

A169. Technical Abstracts Bulletin (TAB). Defense Documentation Center. Department of Defense. v.1-67, 1953-67.

A170. U.S. Government Research and Development Reports. Clearinghouse, v.1, 1946- . (title varies) A600 Un34b
  Many indexes, some privately published, have appeared. Consult: Boylan, Nancy. Identifying Technical Reports Through U.S. Government

Research Reports and Its Published Indexes. <u>Coll. & Res. Lib.</u>
28:175-183, 1967.
  See: Wi-EA74.

A171. U.S. Government Research and Development Reports Index.
Clearinghouse, v.1, 1965-   . (Original title: Government-Wide
Index to Federal Research and Development Reports) 016.6 GOV
  See: Wi-EA75.

A172. Bradshaw, N.H. comp. PB-AD Report Index: BSIR, BTR,
USGRR, USGRDR, 1946-1967. Wash., Technical Information Service,
1968.
  See: <u>Sci-Tech News</u> 22:113, 1968.

A173. CAST - Clearinghouse Announcements in Science and Technology.
Clearinghouse, v.1, 1968- .
  See: <u>Sci. Info. Notes</u> 10:23, (June/July) 1968.

A174. Special Libraries Association. Dictionary of Report Series
Codes. 1st ed., 1962. (2d ed. in preparation) 025.179 S74d
  See: <u>Sp. Libs.</u> 53:574-8, 1962; <u>Am. Doc.</u> 15:59, 1964. Wi-EA60.

A175. U.S. Atomic Energy Commission. Subject Headings used by the
USAEC Division of Technical Information. 1951- . (TID 5001 series)
025.36 U5812s

A176. U.S. Federal Council for Science and Technology. Committee
on Scientific and Technical Information. Standard for Descriptive
Cataloging of Government Scientific and Technical Reports. Revision
No. 1, 1966. Clearinghouse. (AD641 092 or PB173 314) 025.31
U5856S
  See: <u>Sp. Libs.</u> 58:582-4, 1967; <u>Lib. J.</u> 92:183, 1967.

A177. U.S. Library of Congress. Popular Names of U.S. Government
Reports; A Catalog. G.P.O., 1966. 015.73 Un3812
  See: <u>RQ</u> 8:67, 1968.

A178. U.S National Aeronautics and Space Administration. NASA
Thesaurus; Subject Terms for Indexing Scientific and Technical
Information. G.P.O., 1967. 3v. (NASA SP-7030) 025.36 U5868n
  See: <u>Sp. Libs.</u> 59:288, 1968; <u>Am. Doc.</u> 19:208-9, 1968.

<u>PATENTS</u>

A179. Bowker Associates, Inc. United States Patent Previews,
1965-1970; Assignments of Pending Patents Recorded in the U.S.
Patent Office, January 1963-July 1965. 016.608773 B67u
  See: <u>Sp. Libs.</u> 57:667, 1966.

A180. Clavert, R.P. Encyclopedia of Patent Practice and Invention Management. Reinhold, 1964. 608.773 C13e
  See: Chem. & Eng. News 43:56, (Apr. 5) 1965.

A181. Jones, S.V. Inventor's Patent Handbook. Dial, 1967. (Revision of You Ought to Patent That, 1962.) 608.773 J72i
  See: Choice 4:400, 1967.

A182. Newby, F. How to Find Out About Patents, Pergamon, 1967. 608.742 N42h
  See: New Tech. Bks. 52:267-8, 1967.

A183. Patent Licensing Gazette. Techni Research Associates, v.1, 1968- . Bimonthly.

A184. U.S. Patent Office. General Information Concerning Patents. G.P.O., 1922- . 608 Un3ge

A185. U.S. Patent Office. Development and Use of Patent Classification Systems. G.P.O., 1966. 608.773 Un3de

A186. U.S. Patent Office. Manual of Classification of Patents. G.P.O., 1947- . (looseleaf) 608.773 Un3m
  Index to Classification. G.P.O., 1947- . (looseleaf) 608.773 Un3m Index.

A187. U.S. Patent Office. Official Gazette. G.P.O., v.1, 1872- . 608 Un37.
  See: Wi-EA200.

A188. U.S. Patent Office. Official Gazette - Patent Abstract Section. G.P.O., 1967- .
  See: Sci. Info. Notes 9:16-17, (Oct./Nov.) 1967.

A189. U.S. Patent Office. Index of Patents Issued from the U.S. Patent Office. Wash., G.P.O., v.1, 1921- . (Annual) 608 Un37i
  See: Wi-EA202.

See also: Houghton, B. Technical Information Sources: a Guide to Patents, Standards and Technical Reports Literature. Archon, 1967.

For information on foreign patent office publications see: Newby. How to Find Out About Patents. p. 69-99. (No. A182)
  Crane, E.J. et al. A Guide to the Literature of Chemistry. (No. D4)

SERIALS

General

A190. American Society for Testing and Materials. CODEN for Periodical Titles. Philadelphia, A.S.T.M., 1967. 2v. (DS 23A)
Supplements, 1968- . 016.505 W97c
   See: Sp. Libs. 59:279-280, 1968; LRTS 12:359-65, 1968.

A191. Bolton, H.C. Catalogue of Scientific and Technical Periodicals, 1665-1895. 2d ed. Wash., Smithsonian, 1897. (Smithsonian Misc. Coll. v. 40) A505 B63c2
   See: Wi-EA37. Wa p. 29-30.

A192. Fowler, M.J. Guides to Scientific Periodicals: Annotated Bibliography. London, Library Association, 1966. 016.505 F82g
   See: Jour. of Doc. 23:84-5, 1967; Lib. J. 92:2541, 1967.

A193. Scudder, S.H. Catalogue of Scientific Serials of all Countries, Including the Transactions of Learned Societies in the Natural, Physical and Mathematical Sciences, 1633-1876. Harvard Univ. Pr., 1879.
Reprinted by Kraus, 1965. 016.5 Scu2c
   See: Wa p. 31.

A194. Special Libraries Association. Guide to Special Issues and Indexes of Periodicals. N.Y., 1962. 016.05 Sp3g
   See: Sub. Bks. Bull. 59:56, 1962. Wi-AF16.

A195. Ulrich's International Periodical Directory. 12th ed. v. 1: Scientific, Technical, and Medical Periodicals. 1967. Updated between editions by an annual supplement. AO50 P418

A196. World List of Scientific Periodicals Published in the Years 1900-1960. 4th ed., London, Butterworth, 1963-65. 3v. 016.505 W893
Supplemented by British Union-Catalogue of Periodicals, incorporating World List of Scientific Periodicals. New Periodical Titles. 1964-
Quarterly. Annual cumulations. 016.505 W8931
   See: Wi-EA42, AF109. Wa p. 32.

   See also:
      Union List of Serials in the Libraries of the U.S. and Canada.
         3d ed. Wilson, 1965. 5v. q016.05 Un33
      New Serial Titles. Bowker, 1953- . (Monthly, annual cumula-
         tion, five-year cumulation) 016.05 N47
      Kronick, D.A. History of Scientific and Technical Periodicals
         (No. A27)

Special

A197. Australia. Commonwealth Scientific and Industrial Research
Organization. Scientific Serials in Australian Libraries. Melbourne,
v. 1, 1958- . 016.505 Au7c
  See: Wi-EA53. Wa p. 16.

A198. Center for Research Libraries. Rarely Held Scientific Serials
in the Midwest Inter-Library Center. Chicago, 1963. Supplement, 1964.
016.505 M584
  See: Sci. Info Notes 10:22, (June/July) 1968. Wi-EA38. Wa p. 31.

A199. Directory of Canadian Scientific and Technical Periodicals.
Ottawa. National Research Council, 1961-   016.505 D63
  See: L.C. Info. Bull. 26:364, 1967. Wi-Ea43. Wa p. 35.

A200. Guide to Latin American Scientific and Technical Periodicals,
an Annotated List. Wash., Pan American Union, 1962. 016.505 P19g
  See: Wi-EA49. Wa p. 35.

A201. John Crerar Library, Chicago. List of Current Serials.
1965. 016.05 C4321
  See: L.C. Info. Bull. 24:541, 1965; RQ 8:67, 1968.

SCIENTIFIC INFORMATION ACTIVITIES

A202. Annual Review of Information Science and Technology.
Encyclopaedia Britannica, v. 1, 1966- .
  See: Lib. J. 92:1143, 1967; Lib. Q. 37:242-3, 1967; Sp. Libs.
58:431, 1967.

A203. Encyclopedia of U.S. Government Benefits. Wise, 1965.
  See: Lib. J. 90:3268-9, 1965.

A204. Engineers Joint Council. Thesaurus of Engineering and Scientific
Terms. (Project LEX) N.Y., 1967. 025.36 E48th
  See: Mechanical Engineering 91: 50, 1969.

A205. Heald, J.H. The Making of Test Thesaurus of Scientific and
Engineering Terms. Clearinghouse, 1967. (AD 661-001) 025.36 H434m

A206. Klempner, I.M. Diffusion of Abstracting and Indexing Services
for Government Sponsored Research. Scarecrow, 1967.
  See: Lib. J. 92:3529, 1968; Coll. & Res. Libs. 29:518-9, 1968.

A207. Regional Access to Scientific and Technical Information; a Program for Action in the New York Metropolitan Area. Report of the METRO Science Library Project 1966-67. N.Y., METRO New York Metropolitan Reference and Research Library Agency, 1968.
  See: Sci-Tech News 23:22, 1969.

A208. Scientific Information Notes. Wash., National Science Foundation, v.1-10, 1959-68. 500 Sci 285

A208a. Scientific Information Notes. N.Y., Science Associates/ International, v.1, 1969- .
  See: Med. Lib. Assn. Bull. 57:222, 1969.

A209. Special Libraries Association. Mutual Exchange in Scientific Library and Technical Information Fields. N.Y., 1968.
  See: Med. Lib. Assn. Bull. 56:208, 1968.

A210. System Development Corporation. A System Study of Abstracting and Indexing in the United States. Clearinghouse, 1966. (PB 174-249) 029.4 S995s

A211. U.S. Government Organization Manual. G.P.O., 1935- . (title varies) 353 Un343u
  See: Wi-CI55.

A212. U.S. Federal Council for Science and Technology. Committee on Scientific and Technical Information (COSATI). The Copyright Law as It Relates to National Information Systems and National Programs. Clearinghouse, 1967. (PB 175-618) 655.673 Un334c

A213. U.S. Federal Council for Science and Technology. Committee on Scientific and Technical Information (COSATI). Progress of the United States Government in Scientific and Technical Communications, 1966. Clearinghouse, 1967. (PB 176-535)

A214. U.S. Federal Council for Science and Technology. Committee on Scientific and Technical Information (COSATI). Recommendations for National Document Handling Systems in Science and Technology. Clearinghouse, 1965. (PB 168-267 and AD 624-560) 010.78 Un313r
  See: Med. Lib. Assn. Bull. 54:186-8, 1966; Sp. Libs. 57:338-9, 1966.

A215. U.S. Library of Congress. Project MARC: an Experiment in Automating Library of Congress Catalog Data. Wash., 1967. 025.37 U583 pr

A216. U.S. Library of Congress. The MARC II Format: a Communications Format for Bibliographic Data. Wash., 1968. Supplement, 1968- . 025.37 Un583m

A217. U.S. National Bureau of Standards. National Standard Reference Data System: Plan of Operation. G.P.O., 1964. (NSRDS-NBS 1)

A218. U.S. National Science Foundation. Current Research and Development in Scientific Documentation. No. 1, 1957- 010 C937

A219. U.S. National Science Foundation. Federal Organization for Scientific Activities. G.P.O., 1962. (NSF 62-37) 509.73 Un3f

A220. U.S. National Science Foundation. Nonconventional Scientific and Technical Information Systems in Current Use. No. 1, 1956-

A221. U.S. National Science Foundation. Scientific Information Activities of Federal Agencies. No. 1, 1959- . (series of pamphlets) 015.732 Un36s

A222. U.S. Office of Naval Research. Manual for Building a Technical Thesaurus. 1966. 025.36 Un58m

A223. U.S. President's Science Advisory Committee. Science Government and Information: The Responsibility of the Technical Community and the Government in the Transfer of Information. (Weinberg Report). G.P.O., 1963. 501 Un3s

A224. Warren, S.L. Proposal: The National Library of Science System and Network for the Published Scientific Literature. Wash. Office of the Special Assistant to the President for Mental Retardation. 1964.

See also the units on DIRECTORIES, PATENTS, and RESEARCH AND DEVELOPMENT REPORTS.

CURRENT INFORMATION; SELECTION AIDS

See: Chemical & Engineering News. (No. D164)

A225. FID News Bulletin. The Hague, International Federation for Documentation, 1951- . (Title varies) 016.8205 FI

A226.   Nature.   London, Macmillan, v.1, 1869-   505N

A227.   Science.   Am. Assn. Adv. Sci., v.1, 1883-

A228.   Science Books: a Quarterly Reveiw.   Am. Assn. Adv. Sci.,
v.1, 1965-   .   016.505 Sc261
  See: Lib. J. 91:676-7, 1966.

A229.   Science News.   Science Service, v.1, 1921-   505 ACIN

A230.   Scientific American.   N.Y., v.1, 1845-   605 SA

A231.   Scientific Research.   McGraw-Hill, v.1, 1966-
  See also:
      Choice. v.1, 1964-   .   Choice-on-Cards.   1968-   .   028.05 CHO
      Library Journal. v.1, 1876-   .   (Particularly the special issues
        on reference aids and science books) 020.5 LJ
      New York Public Library.   New Technical Books. v.1, 1915-
        028.66 N532n
      Off the Press; the New Books in Science, Technology, Business,
        and Medicine from All Publishers. Palo Alto, Stacey, 1967-

SECTION B. MATHEMATICS

Mathematics is the logical study of numerical quantities and their relations, of spatial quantities and the relationships between them, and of various abstractions of these relationships. There are three major approaches: pure or theoretical mathematics, applied mathematics, and interpretive mathematics or statistics. The science is divided, at least for an elementary approach, into large subfields as:

Arithmetic: study of the properties and relationships of numbers and of computation with them.

Algebra: generalized arithmetic which uses symbols in place of numbers.

Geometry: deals with space and its relationships; the study of mutual relationships of points, lines, angles, surfaces and solids.

Analysis: investigation of the relationships of variables or indeterminate quantities by means of symbols.

GUIDES TO THE LITERATURE

B1. Goldman, S. Guide to the Literature of Engineering, Mathematics, and the Physical Sciences. 2d ed. John Hopkins University Applied Physics Laboratory, 1964. 016.5 G56g
See: Wa p. 6 (note)

B2. Parke, N.G. Guide to the Literature of Mathematics and Physics Including Related Works on Engineering Science. 2d ed. Dover, 1958. 016.53 P22g
See: Sci. Ref. Notes 6:4, (Jan.) 1959. Wi-EF1. Wa p. 53.

B3. Pemberton, J.E. How to Find Out in Mathematics. Pergamon, 1963; Macmillan, 1964. 016.51 P36h
See: Science 144:831, 1964; Math. of Comp. 18:676, 1964. Wa p. 54.

B4. Schaaf, W.L. Recreational Mathematics: a Guide to the Literature. Wash., National Council of Teachers of Mathematics, 1963. 016.51087 Schlr
See: Rec. Math. Mag. No. 14, p. 30, 1964.

See also: Lohwater, A.J. Mathematical Literature. In Downs, R.B. & Jenkins, F.B., eds. Bibliography. p. 514-529. (No. A3)

BIBLIOGRAPHIES, INDEXES, ABSTRACTS

B5. American Mathematical Society. Contents of Contemporary Mathematical Journals. 1969- . 016.51 C767

B6. Bulletin des Sciences Mathématiques. v.1, 1870- . 510.5 BU

B7. Forsythe, G. Bibliography of Russian Mathematics Books. Chelsea, 1956. 016.51 F77b
  See: Math. Rev. 18:550, 1957; Math. of Comp. 12:244-6, 1958. Wi-EF4.

B8. Jahrbuch über die Fortschritte der Mathematik. Berlin, De Gruyer, v.1, 1868-1947. (In 1934, Revue Semestrielle des Publications Mathematiques, v.1-39, 1893-1934 merged with the Jahrbuch). 510.5 JA
  See: Wi-EF10. Wa p. 53 (note).

B9 Karpinski, L.C. Bibliography of Mathematical Works Printed in America through 1850. Univ. of Mich. Pr., 1940. A510 K14b
  See: Wi-EF6.

B10. Mathematical Reviews. Amer. Math. Soc., v.1, 1940- . 510.5 MATR
  See: Wi-EF11. Wa p. 53.

B11. Smith, D.E. Rara Arithmetica; a Catalogue of the Arithmetics Written before the Year 1601 Ginn, 1908. Addenda, 1939. 016.511 Sm5r
  See: Wi-EF8.

B12. Zentralblatt für Mathematik und ihre Grenzgebiete. Berlin, Springer, v.1, 1931- . (Suspended Nov. 1944-June 1948) 510.5 ZEN
  See: Wi-EF13. Wa p. 53 (note)

  See also:
    Bulletin Signalétique. 1: Mathématiques, pures et appliquées.
      (No. A18)
    International Catalogue of Scientific Literature. A: Mathematics
      (No. A16)
    Referativnyi Zhurnal. Matematika (Bibliographical Index, 1953-
      63. Moscow, 1966. 3v.) (No. A19)

REVIEWS & SURVEYS

B13. Advances in Mathematics. Academic, v.1, 1961-   510 Ad95

B14. Annals of Mathematics. Princeton Univ. Pr., v.1, 1884-
510.5 ANA

B15. National Academy of Sciences. The Mathematical Sciences: A
Report. Wash., 1968.
  See: Chem. & Eng. News 46:16, (Dec. 2) 1968.

HISTORIES

B16. Boyer, C.B. A History of Mathematics. Wiley, 1968.
  See: Am. Scientist 56:468A-9A, 1968; Science 163:171, 1969.

B17. Cajori, F. A History of Mathematical Notations.   London,
Open Court Co., 1928-29. 2v. 510.9 C12his
  See: Wa p. 57.

B18. Midonick, H.O. The Treasury of Mathematics. A Collection of
Source Material in Mathematics ... Philosophical, 1965.
  See: Science 147:1437, 1965.

B19. Miller, G.A. Historical Introduction to Mathematical Literature.
Macmillan, 1916. 510.9 M61h
  See: Wa p. 53 (note).

B20. Montucla, J.E. & Lalande, J.J.L. Histoire des Mathématiques.
2d ed. Paris, Agasse, 1799-1802. 4v. (Reprinted in 1960) 510.9
M76h
  See: S-H Book News 15:6, (Sept.) 1960.

B21. Sarton, G.A. The Study of the History of Mathematics. Harvard
Univ. Pr., 1936. (Reprinted by Dover, 1960) 510.9 Sa7s
  See: New Tech. Bks. 21:31, 1936. Wa p. 57.

B22. Scott, J.F. History of Mathematics. London, Taylor & Francis,
1958. 510.9 Sco8h
  See: Science 128:469, 1958; Sci. Am. 199:141-5, (Oct.) 1958. Was p. 57.

B23. Smith, D.E. History of Mathematics. Ginn, 1923-5. 2v. (Reprinted by Dover, 1958) 510.9 Sm5hi

31

B24. Smith, D.E. Source Book of Mathematics. McGraw-Hill, 1929.
510.9 St8c

B25. Struik, D.J. A Concise History of Mathematics. 3d rev. ed.
Dover, 1967. 510.9 St8c
  See: New Tech. Bks. 52:148-9, 1967.

B26. Struik, D.J., ed. A Source Book in Mathematics, 1200-1800.
Harvard Univ. Pr., 1969.
  See: Science News 95:148, 1969. New Tech. Bks. 54:116, 1969.

DICTIONARIES & ENCYCLOPEDIAS

General

B27. Aleksandrov, A.D. et al. Mathematics: Its Content. Methods
and Meaning. 2d ed. MIT Pr., 1969. 3v. 510 Akl14Eg

B28. Baker, C.C.T. Dictionary of Mathematics. Hart, 1966. 510.3
B17d
  See: Sp. Libs. 57:417, 1966.

B29. Dictionary of Physics and Mathematics Abbreviations, Signs,
and Symbols. Odyssey, 1965. 530.148 D561

B30. Encyklopädie der mathematischen Wissenschaften mit Einschluss
ihrer Anwendungen. Leipzig, Teubner, 1898-1935. 6v. in 23. Bd. 1,
Arithmetik und algebra (2v.); Bd. 2, Analysis (5v.); Bd. 3, Geometrie
(6v.); Bd. 4, Mechanik (4v.); Bd. 5, Physik (3v.); Bd. 6, t. 1, Geodäsie
und Geophysik (1v.); Bd. 6, t. 2, Astronomie (2v.). (Issued also in an
incomplete, somewhat revised, French edition: Encyclopédie des Sciences
Mathematiques Pures et Appliquées, Paris, Gautier-Villars, 1904-16. 7v.)
510.3 Enl
  See: Wi-EF14.

B31. Hogben, L. Mathematics for the Million. 4th ed. Norton, 1969.
510 H67m

B32. International Dictionary of Applied Mathematics. Van Nostrand,
1960. (Multilingual indexes) 510.3 In8
  See: Sci. Ref. Notes 8:12, (Jan.) 1961; Math. of Comp. 16:505-6,
1962. Wi-EF15. Wa p. 55.

B33. James, G. & James, R.C. eds. James & James Mathematics
Dictionary. 3d ed. Van Nostrand, 1968. 510.3 J23m
  See: Lib. J. 93:2847, 1968.

B34. Karush, W. The Crescent Dictionary of Mathematics. Macmillan, 1962. 510.3 K14c
  See: Math. of Comp. 17:478, 1963. Wi-EF17. Wa p. 55 (note).

B35. Millington, T.A. & Millington, W. Dictionary of Mathematics. London, Cassell, 1966. 510.3 M62d
  See: Wi-1EF3.

B36. Naas, J. & Schmid, H.L. Mathematisches Wörterbuch mit Einbeziehung der Theoretischen Physik. Pergamon, 1961. 2v.
  See: Nature 196:608, 1962. Wi-EF18. Wa p. 54-5.

B37. Newman, J.R. ed. The World of Mathematics; a Small Library of the Literature of Mathematics ... Simon & Schuster, 1956. 4v. 510.4 N46w
  See: Sci. Am. 195:147, (Nov.)1956. Wa p. 57 (note).

B38. Séminaire Bourbaki. N.Y., Benjamin. 1948/49-1964/65. 12v. (Additional volumes published annually) 513.06 PAR
  See: New Tech. Bks 51:292, 1966; Science 154:1440, 1966.

B39. Universal Encyclopedia of Mathematics. Simon & Schuster, 1964. (Based on Meyer's Rechenduden published in 1960 by Bibliographisches Institut, Mannheim.) 510.3 M576E
  See: Math. of Comp. 19:164-5, 1965. Wi-EF19. Wa p. 55.

Note discussion of mathematical concepts in general encyclopedias.

Foreign language

B40. Herland, L.J. Dictionary of Mathematical Sciences. 2d ed. rev. London, Harrap, 1965-66. 2v. v.1: German-English. v. 2: English-German. 510.3 H42d
  See: Bibl. Doc. Term. 7:53, (Mar.) 1967. Wi-1EF2.

B41. Macintyre, S. & Witte, E. German-English Mathematical Vocabulary. 2d ed. Interscience, 1966. 510.3 M18g
  See: New Tech. Bks. 51:185, 1966.

B42. Burlak, J. Russian-English Mathematical Vocabulary. Interscience, 1963. 510.3 B87r
  See: New Tech. Bks. 51:210, 1966. Wa p. 56 (note).

B43. Kotz, S. Russian-English Dictionary and Reader in the Cybernetical Sciences. Academic, 1966. 006 K849r

B44. Milne-Thomson, L.M. Russian-English Mathematical Dictionary.
Univ. of Wis. Pr., 1962. 510.3 M63r
  See: Wi-EF22. Wa p. 56.

B45. Russian-English Dictionary of the Mathematical Sciences. Am.
Math. Soc., 1961. 510.3 R92
  See: Wi-EF23.

HANDBOOKS & TABLES

Guides to tables

B46. Fletcher, A. et al. An Index of Mathematical Tables. 2d ed.
Addison-Wesley, 1962. 510.8 F63i
  See: Science 137:332, 1962. Wi-EF32. Wa p. 59.

B47. Lebedev, A.V. & Fedrova, R.M. A Guide to Mathematical
Tables. English edition. Pergamon, 1960. Supplements, 1960-   .
016.5108 L49sEf
  See: Math. of Comp. 11:104-6, 1957 (Russian edition). Wi-EF34.
Wa p. 59.

B48. Schütte, K. Index Mathematischer Tafelwerke und Tabellen.
(Index of Mathematical Tables). 2d ed. Munich, Oldenbourg, 1966.
A510.8 Sch8i
  See: Math. of Comp. 21:501-2, 1967.

B49. Mathematics of Computation. v. 1, 1943-   .   (title varies)
510.5 MATA
  Cumulative subject/author index in preparation. See: Sci. Info.
Notes 10:23, (April/May) 1968.

Tables

B50. Abramowitz, M. & Stegun, I.A. eds. Handbook of Mathematical
Functions, with Formulas, Graphs, and Mathematical Tables. G.P.O.,
1964. (Applied Mathematics Series, No. 55) Paperbound, 1965.
517.35 Ab83h
  See: Math. of Comp. 19:147-9, 1965; Am. Scientist 52:456A, 1964.
Wi-EF37. Wa p. 58.

B51. Burington, R.S. Handbook of Mathematical Tables and Formulas. 4th ed. McGraw-Hill, 1965. 510.8 B92h
  See: Math. of Comp. 19:503, 1965. Wi-EF42. Wa p. 58.

B52. CRC Standard Mathematical Tables. 15th ed. Chem. Rubber Co., 1965. 510.8 M432

B53. Great Britain. D.S.I.R. National Physical Laboratory. Mathematical Tables. London, H.M.S. Office, 1956-
  See: Wa p. 59.

B54. Handbook of Mathematical Tables. Supplement to Handbook of Chemistry and Physics. Chem. Rubber Co., 1962- . 510.8 H192
  See: Math. of Comp. 19:680, 1965. Wa p. 59.

B55. Jansson, M.E. et al. Handbook of Applied Mathematics. 4th ed. Van Nostrand, 1966. 510 J26h
  See: Wi-1EF4.

B56. Korn, G.A. & Korn, T.M. Mathematical Handbook for Scientists and Engineers; Definitions, Theorems, and Formulas for Reference and Review. 2d rev. ed. McGraw-Hill, 1968. 510 K842m
  See: New Tech. Bks. 53:78, 1968.

B57. Royal Society. Royal Society Mathematical Tables. Cambridge Univ. Pr., 1950- . (11v. in progress) 510.8 R81r
  See: Wi-EF48. Wa p. 59-60.

   Note: Handbooks of chemistry, physics, and engineering contain mathematical tables.

DIRECTORIES

B58. World Directory of Mathematicians. International Mathematical Union, 1958- . 510.97 W893
  See: Wi-EF25. Wa p. 56.

COMPUTERS

Bibliographic aids

B59. Advances in Computers. Academic, v.1, 1960- . 510.84 Ad95
  See: Wa p. 410.

B60. Computer Abstracts. London, Technical Information Co., v. 1,
1960-   . 510. 8405 COA
  See: Wa p. 410.

B61. Computer Literature Bibliography. Wash., G.P.O., 1965-   .
(v. 1: 1946-63; v. 2: 1964-67) 016. 51392 Y83c
  See: Sci. Info. Notes 7:14, (June/July) 1965. Wa p. 410.

B62. International Journal of Computer Mathematics. N.Y., Gordon &
Breach. v. 1, 1964-   . 510. 8405 INT
  See: New Tech. Bks. 49:296, 1964.

Dictionaries

B62a. Condensed Computer Encyclopedia. McGraw-Hill, 1969.
  See: Science News 95:491, 1969.

B63. Horn, J. Computer and Data Processing: Dictionary and Guide.
Prentice-Hall, 1966. 651. 803 H78c
  See: New Tech. Bks. 52:175, 1967.

B64. Sippl, C.J. Computer Dictionary. Sams, 1966. 651.2603 Si7c
  See: New Tech. Bks. 51:151, 1966; Sp. Libs. 57:669, 1966.

B65. Spencer, D.D. The Computer Programmer's Dictionary and
Handbook. Ginn, 1968.
  See: Lib. J. 93:3545, 1968.

B66. Trollham, L. & Whittmann, A. Dictionary of Data Processing.
Elsevier, 1965. (Terms in English/American, German, French)
651.2603 T74d
  See: Math. of Comp. 19:703-4, 1965. Wa p. 411.

B67. Weik, M. Dictionary of Computer Terms. Hayden, 1968.
  See: Lib. J. 93:2698, 1968.

Handbooks & Yearbooks

B68. Auerbach Computer Notebook. Philadelphia, Auerbach Info,
Inc., 1968-   . (Monthly) q651. 8 Au3a
  See: Bibl. Doc. Term. 8:171, 1968.

B69. Computer Graphics: Techniques and Applications. Plenum, 1969.

B70. Computer Yearbook and Directory. Detroit, American Data Processing; 1966. 510.84 C74

B71. Klerer, M. & Korn, G.A. Digital Computer User's Handbook. McGraw-Hill, 1967. 510.84 K67d
  See: Lib. J. 92:3626, 1967.

STATISTICS

Bibliographic aids

B72. International Statistical Institute. Bibliography of Basic Texts and Monographs on Statistical Methods, 1945-1960. 2d ed. Edinburgh, Oliver & Boyd, 1963. 016.3112 In8b
  See: Wa p. 60

B73. Kendall, M.G. & Doig, A.G. Bibliography of Statistical Literature. Edinburgh, Oliver & Boyd, 1962- . (v.1 indexes the literature of 1950-1958; v.2 indexes the literature of 1940-1949; a third volume is planned to cover the literature of the 16th century to 1939) 016.311 K33b
  See: Wa p. 61.

B74. Lancaster, H.O. Bibliography of Statistical Bibliographies. Benjamin, 1968.
  See: Lib. J. 93:2698, 1968.

B75. Statistical Theory and Method Abstracts. Edinburgh, Oliver & Boyd, v.1, 1959. (With v.5, 1964 merged with International Journal of Abstracts: Statistical Theory and Method, v. 1-10, 1954-63) 311.05 INJ
  See: Wa p. 61.

Dictionaries

B76. Hungarian Central Statistical Office. Statistical Dictionary; 1700 Statistical Terms in Seven Languages. Budapest, 1961. 311.03 H895.
  See: Wi-CG16.

B77. Kendall, M.G. & Buckland, W.R. A Dictionary of Statistical Terms. 2d ed. Hafner, 1960. (Glossaries in French, German, Italian, and Spanish) 311.03 K33d
  See: Wi-CG18. Wa p. 61.

B78. Kotz, S & Hoeffding, W. Russian English Dictionary of Statistical Terms and Expressions and Russian Reader in Statistics. Univ. of N.C. Pr., 1964.
  See: Science 147:143, 1965. Wa p. 61 (note).

B79. Kurtz, A.K. & Edgerton, H.A. Statistical Dictionary of Terms and Symbols. Wiley, 1939. Reprinted by Hafner, 1967.
  See: Choice 5:608, 1968. Wi-CG19.

Handbooks

B80. Greenwood, J.A. & Hartley, H.O. Guide to Tables in Mathematical Statistics. Princeton Univ. Pr., 1962. 016.51912 G85g
  See: Math. of Comp. 18:157-8, 1964; Science 137:332, 1962.
Wi-EF33. Wa p. 61.

B81. Bancroft, T.A. Topics in Intermediate Statistical Methods. Iowa State Univ. Pr., v. 1, 1968- .
  See: Choice 5:1468-9, 1969.

B82. Beyer, W.H. CRC Handbook of Tables for Probability and Statistics. Cleveland, Chemical Rubber Co., 1966. 519 B467c
  See: Science 154:1316, 1966. Wi-EF5.

B83. Chakravarti, I.M. et al. Handbook of Methods of Applied Statistics. Wiley, 1967. 2v. 519 C34h
  See: Choice 4:460, 1968.

B84. Johnson, N.L. & Leone, F.C. Statistics and Experimental Design in Engineering and the Physical Sciences. Wiley, 1964. 2v.
519.9 J63s
  See: Science 147:1025, 1965.

B85. Neville, A.M. & Kennedy, J.B. Basic Statistical Methods for Engineers and Scientists. International Textbook, 1964. 311.2 N41b
  See: New Tech. Bks. 50:71, 1965.

B86. Owen, D.B. Handbook of Statistical Tables. Addison-Wesley, 1962. 310.83 Ow2h
  See: New Tech. Bks. 48:39, 1963. Wi-EF47. Wa p. 61.

B87. Walker, H.M. Mathematics Essential for Elementary Statistics. Rev. ed. Holt, 1951. 311 W15m

CURRENT INFORMATION; SELECTION AIDS

Consult:

American Mathematical Society. New Publications.

Mathematical Association of America. Committee on the
   Undergraduate Program on Mathematics. Basic Library
   List. Berkeley, Calif., 1963.

Current issues of:

The American Mathematical Monthly, v. 1, 1894- .
American Mathematical Society. Bulletin, v. 1, 1894-   .
Journal of Recreational Mathematics, v. 1, 1968-   .
Mathematical Reviews, v. 1, 1940-   .
Mathematics of Computation (extensive section: Reviews and
   Descriptions of Tables and Books), v. 1, 1943-   .
The Mathematics Teacher, v. 1, 1908-   .
School Science and Mathematics, v. 1, 1901-   .
Scripta Mathematica, v. 1, 1933-   .

SECTION C. PHYSICS

Physics is the science or group of sciences which deals with matter and energy and the relationships between them. Physics is divided into classical physics and modern physics. These disciplines in turn are subdivided into such areas as:

Classical physics:
> Mechanics: concerned with the effects of force acting upon bodies at rest and in motion.
> Heat: deals with the nature of temperature changes.
> Acoustics: that division which deals with the study of sound.
> Optics: concerned with the study of visible and invisible light under varied conditions.
> Electricity and magnetism: study of the principles of electric currents and circuits, the earth's magnetic field and the power of electromagnets.

Modern physics:
> Atomic physics: study of atoms and their structure.
> Nuclear physics: deals with the study of the nucleus of the atom and with the energy produced when nuclear particles are disturbed by force.
> Electronics: concerned with the study and control of the movement of free electrons, including the applications involving ions.
> Quantum optics: studies involving the quantum theory of light.
> Plasma physics: study of a plasma, a hot gas composed of ions and electrons which conduct electricity.
> Solid state physics: study of the properties and structures of solid materials.

GUIDES TO THE LITERATURE

C1. Anthony, L. J. Sources of Information on Atomic Energy. Pergamon, 1966. 016.5397 An8s

C2. Maizell, R. E. & Seigel, F. The Periodical Literature of Physics. N. Y., Am. Inst. Physics, 1961. 016.5305 M28p

C3. U. S. Atomic Energy Commission. Technical Books and Monographs. 1st ed., 1959- .016.539 Un32t
  See: Chem. & Eng. News 46:68, (Aug. 5) 1968.

C4. U.S. Atomic Energy Commission. What's Available in the Atomic Energy Literature. 1955- . (TID-4550 revised frequently) 016.539 Un32w

C5. Whitford, R.H. Physics Literature: A Reference Manual. 2d ed., Scarecrow, 1968. A530 W58p
  See: New Tech. Bks. 54:47, 1969; Choice 5:1432, 1969.

C6. Yates, B. How to Find Out About Physics. Pergamon, 1965. 016.53 Y27h
  See: Sp. Libs. 57:204, 1966.

  See also: Parke, N.G. Guide to the Literature of Mathematics and
      Physics. (No. B2)

BIBLIOGRAPHIES, INDEXES, ABSTRACTS

General

C7. Current Papers in Physics (CCP). London, Institute of Electrical Engineers, v.1, 1966- . 016.5305 CP
  See: Sci. Info. Notes 8:17, (Dec./Jan.) 1965-66.

C8. Physics Abstracts (Science Abstracts, Section A). London, Institute of Electrical Engineers, v.1, 1898- . 530.5 Sa
  See: Wi-EG12. Wa p. 72.

C9. Physics Express. N.Y., International Physics Index, v. 1, 1958- 530.5 PHYE
  See: Wa p. 72.

C10. Physikalische Berichte. Braunschweig, Germany, Vieweg, v.1, 1920- . (Continues Fortschritte der Physik, 1845-1918) 530.5 PHB
  See: Wi-EG11. Wa p. 72.

C11. Markworth, M.L. Dissertations in Physics: an Indexed Bibliography to All Doctoral Theses Accepted by American Universities, 1861-1959. Stanford Univ. Pr., 1961. q016.53 M33d

  See also:
    Bulletin Signalétique. 3-4: Physique. (No. A18)
    International Catalogue of Scientific Literature. C: Physics.
      (No. A16)
    Referativnyi Zhurnal. Fizika. (No. A19)

Special

C12. International Atomic Energy Agency. List of Bibliographies on Nuclear Energy. Vienna, v. 1, 1960- (Irregular) 016. 5397 In8li

C13. Rheology Abstracts: a Survey of World Literature. Pergamon, v. 1, 1958- . 530. 0305 RH
  See: Wa p. 85.

C14. Solid State Abstracts. Cambridge, Mass., Cambridge Communications Corp., v. 1, 1960- . 539. 105 SOL
  See: Wa p. 84.

C15. United Nations. International Bibliography on Atomic Energy. Columbia Univ. Pr., 1949-51. 2v. Supplements, 1950-1953. (v. 1; Political, economic and social aspects, 1949. Supplements, 1950, 1953. v. 2: Scientific aspects, 1949-1951. Supplements, 1952, 1953) qA539 Un311i
  See: Wi-EG9.

C16. U. S. Atomic Energy Commission. Bibliographies of Interest to the Atomic Program. Oak Ridge, 1958- . (TID 3043 series) 539 Un32t 3043
  See: Wi-EI201. Wa p. 218.

  See also:
    Nuclear Science Abstracts. (No. A167)
    Mathematical Reviews. (No. B10)
    Chemical Abstracts. (No. D14)

REVIEWS & SURVEYS

C17. Advances in Atomic and Molecular Physics. Academic, v. 1, 1965- 539 Ad9

C18. Advances in Chemical Physics. Interscience, v. 1, 1958- 530 Ad95.

C19. Advances in Electronics and Electron Physics. Academic, v. 1, 1948- . Supplements, 1963- . 621. 36 Ad95
  See: Wa p. 234.

C20. Advances in Nuclear Physics. Plenum, v. 1, 1968- 539. 7 Ad9
  See: Science 160:1331, 1968.

C21. **Advances in Plasma Physics.** Interscience, v. 1, 1968-
537. 532 Ad9
  See: Science 163:803, 1969; Choice 6:250, 1969.

C22. **Advances in Theoretical Physics.** Academic, v. 1, 1965-
530 Ad96

C23. Annual Review of Nuclear Science. Annual Review Inc., v. 1,
1952- . 539 An78

C24. Comments on Nuclear and Particle Physics. Gordon & Breach,
v. 1, 1967. 539. 105 CO

C25. Comments on Solid State Physics. Gordon & Breach. v. 1,
1968- .

C26. Progress in Physics. London, Physical Society, v. 1, 1934-
(Formerly Reports on Progress in Physics, v. 1-31, 1934-68).
530. 4 P6r

C27. Reviews of Modern Physics. American Institute of Physics,
v. 1, 1929- . 530. 5 REV

C28. Solid State Physics: Advances in Research and Application.
Academic, v. 1, 1955- . Supplements, 1958- . 539. 1 So4
  See: Wa p. 84.

HISTORIES

C29. Auerbach, F. Geschichtstafeln der Physik. Leipzig, Barth,
1910. 530. 9 Au3g

C30. Cajori, F. A History of Physics. Rev. ed., Macmillan, 1929.
530. 9 C12h

C31. Chalmers, T. W. Historic Researches; Chapters in the History
of Physical and Chemical Discovery. Scribner, 1952. 530. 9 C35h

C32. Chase, C. T. The Evolution of Modern Physics. Van Nostrand,
1947. 530. 9 C38e

C33. Crew, H. The Rise of Modern Physics. 2d ed. Williams &
Wilkins, 1935. 530. 9 C86r2

C34. Glasstone, S. Sourcebook on Atomic Energy. 3d ed. Van
Nostrand, 1967. 539 G46s
  See: New Tech. Bks. 53:10, 1968; Choice 5:181, 1968.

43

C35. Magie, W. F. Source Book in Physics. McGraw-Hill, 1935.
530.9 M27s

C36. Nobel Lectures in Physics. Am. Elsevier, 1964-67. 3v.
(v.1, 1901-1921; v.2, 1922-1941; v.3, 1942-62) 530.08 N66p
  See: Science 156:776-7, 1967.

C37. Taylor, L. W. Physics, the Pioneer Science. Houghton, 1941.
530.01 T21p

DICTIONARIES

  General

C38. Gray, H. J. ed. Dictionary of Physics. Longmans, 1958.
530.3 G79d
  See: New Tech. Bks. 43:67-8, 1958. Wi-EG16. Wa p. 74.

C39. Hix, C. F. & Alley, R. P. Physical Laws and Effects. Wiley,
1958. 530.02 H64p
  See: Sci. Ref. Notes 6:22, 1959. Wi-EG35. Wa p. 76.

C40. International Dictionary of Physics and Electronics. 2d ed.
Van Nostrand, 1961. 530.3 In8
  See: Sci. Ref. Notes 8:21, (Apr./Oct.) 1961.

  See also:
  Dictionary of Physics and Mathematics Abbreviations. (No. B29)

  Foreign language

C41. Carpovich, E. A. Russian-English Atomic Dictionary. 2d ed.
rev. Technical Dictionaries, 1959. 539.03 K14r
  See: Chem. & Eng. News 37:86+, (Sept.) 1959. Wi-EI208. Wa p. 220.

C42. DeVries, L. & Clason, W. E Dictionary of Pure and Applied
Physics. Elsevier, 1963-64. 2v. v.1: German-English. v. 2:
English-German. 530.3 D49d
  See: Science 143:1426-7, 1964. Wi-EG25. Wa p. 75.

C43. Dictionary of Physics and Allied Sciences. Ungar, 1958-1962.
2v. v. 1: German-English: v.2: English-German. 530.3 D561
  See: Wi-EG26. Wa p. 75.

44

C44. Emin, I. et al. Russian-English Physics Dictionary. Wiley, 1963. 530.3 Em47r
  See: Science 143:121-2, 1964. Wi-EG31. Wa p. 75.

C45. Lettenmeyer, L. Dictionary of Atomic Terminology. Philosophical, 1959. (English, German, French, Italian) 539.703 L56
  See: Sci. Ref. Notes 6:23, (July) 1959. Wi-EG29.

C46. Rau, H. Dictionary of Nuclear Physics and Nuclear Chemistry. 2d ed. London, Pitman, 1965. (German-English/American, English/American-German) 539.703 R19w
  See: Wa p. 220.

ENCYCLOPEDIAS & COMPREHENSIVE WORKS

  General

C47. Asimov, I. Understanding Physics. Walker, 1967. 3v. 530 As4u.
  See: New Tech. Bks. 52:80, 1967.

C48. Besançon, R.M. ed. The Encyclopedia of Physics. Reinhold, 1966. 530.3 B46e
  See: Physics Today 19:97, (Oct.) 1966; Science 152:951, 1966.

C49. Concise Encyclopaedia of Nuclear Energy. Interscience, 1963. 539.7603 C76
  See: Chem. & Eng. News 41:70, (June 17) 1963.

C50. Encyclopaedic Dictionary of Physics: General, Nuclear, Solid State, Molecular, Chemical, Metal and Vacuum Physics, Astronomy, Geophysics, Biophysics, and Related Subjects. Macmillan, 1961-65. 9v. (v. 9, a multilingual glossary, is available separately) Supplements, 1966- . 530.3 En192
  See: Sub. Bks. Bull. 64:1061-64, 1968; Choice 4:636, 1967. Wi-EG14 and IEG3. Wa p. 73-4.

C51. Handbuch der Physik. 2d ed. Berlin, Springer, 1955- . (To be completed in 54 volumes in approximately 70 parts; articles in English, French and German.) 530 H195
  See: Wi-EG18 and 1EG5. Wa p. 73.

C52. Methods of Experimental Physics. Academic, 1959-   (v.1:
Classical methods; v.2: Electronic methods; v.3: Molecular physics;
v.4: Atomic and electron physics; v.5: Nuclear physics; v.6: Solid
state physics) 530 M367m
   See: Physics Today 21:84-85, (June) 1968.

   Special

C53. Burhop, E.H.S. ed. High Energy Physics. Academic. v.1, 1967-
(3v. projected) 539.72108 B91h
   See: Physics Today 21:87-9, (July) 1968; Science 158:251, 1967.

C54. Clark, G.L. ed. Encyclopedia of X-rays and Gamma Rays.
Reinhold, 1963. 537.535203 C54e
   See: New Tech. Bks. 49:5, 1964. Wi-EG24. Wa p. 83.

C54a. Cochran, J.F. & Haering, R.R. eds. Solid State Physics.
Gordon & Breach, 1968-   .
   See: New Tech. Bks. 54:85, 1969.

C55. Hogerton, J.F. The Atomic Energy Deskbook. Reinhold, 1963.
539.7603 H679a
   See: Chem. & Eng News 42:43-4, (Jan. 20) 1964. Wi-EG20.
Wa p. 222.

C56. International Encyclopedia of Physical Chemistry and Chemical
Physics. Pergamon, 1960-   (A series of monographs planned in
about 100 volumes)
   See: Wa p. 97.

C57. Mason, W.P. ed. Physical Acoustics. Academic, 1964-68.
534 M38p
   See: Physics Today 21:85-6, (Feb.) 1968.

HANDBOOKS & TABLES

   See also sections GENERAL SCIENCE, MATHEMATICS, CHEMISTRY
and ENGINEERING SCIENCES

   General

C58. American Institute of Physics Handbook. 2d ed. McGraw-Hill,
1963. 530.02 Am3a
   See: Sci. Ref. Notes 10:24, 1963. Wi-EG33. Wa p. 76.

C59. Condon, E.U. & Odishaw, H. eds. **Handbook of Physics.** 2d ed.
McGraw-Hill, 1967. 530 C75h
　See: Science 160:1440, 1968; Choice 5:325, **1968.**

C60. Ebert, H. ed. Physics Pocketbook. **Interscience,** 1968.530.8
Eb 3pE
　See: New Tech. Bks. 53:10, 1968; Am. Scientist 56:275A, 1968.

C61. Menzel, D.H. Fundamental Formulas of Physics. Prentice-
Hall, 1955. Dover, 1960. 530.15 M52f
　See: New Tech. Bks. 40:66, 1955. Wi-EG39, Wa p. 76.

　　Special

C62. Frisch, O.R. ed. Nuclear Handbook. Van Nostrand, 1958.
539.5 F917n
　See: Sci. Ref. Notes 6:20, (Apr.) 1959. Wi-EI210. Wa p. 222.

C63. Kunz, W. & Schintlmeister, J. Nuclear Tables. Pergamon,
1958-　. q539.7 K964n

C64. Nuclear Data. Academic. Section A, 1965-　. Section B,
1966-　. (Supersedes National Research Council's Nuclear Data
Sheets) 539.705 NUC

C65. Nuclear Tables. Pergamon, 1958-　. q539.7 K964r

C66. Nuclear Theory Reference Book. National Research Council,
1959-　. (Biennial compilation of Nuclear Theory Index Cards,
1958-　.) 539.7 N213nu

C67. Kaelbe, E.F. ed. Handbook of X-Rays. McGraw-Hill, 1967
539.722 K11h
　See: New Tech. Bks. 53:47, 1968.

C68. Maerz, A. & Paul, M.R. A Dictionary of Color. 2d ed.,
McGraw-Hill, 1950. q752 M26d
　See: Wi-EG47.

C69. U.S. National Bureau of Standards. The ISCC-NBS Method
of Designating Color Names. G.P.O., 1955. (National Bureau of
Standards Circular 553) 752 Un3i
　See: Science 123:678, 1956.

47

C70. Wyszecki, G. & Stiles, V.S. **Color Science.** Wiley, 1967.
535.6 W99c
   See: Physics Today 21:83-4, **(July)** 1968.

   See also: Chapanis, A. **Color Names for Color Space.** Am.
Scientist 53:327-46, 1965.

INFORMATION ACTIVITIES

C71. Alt, F.L. Plans for a National Physics Information System.
N.Y., American Institute of Physics, 1968. 010.78 A179p

C72. Herschman, A. Information Retrieval in Physics. N.Y.,
American Institute of Physics, 1967. (IARD 67-1)

DIRECTORIES

C73. Atomic Handbook. London, Morgan. v.1: Europe. 1965-
539.76 At71

C74. Who's Who in Atoms. London, Vallency, 1949-   925.3 W62
   See: Wi-EA196. Wa p. 223.

C75. World Nuclear Directory. London, Harrap Research Labs.,
1960 -   539.7058 W89

SERIALS

   See: Maizell, R.E. & Seigel, F. The Periodical Literature of
Physics. (No. C2)

C76. Keenan, S. & Atherton, P. The Journal Literature of Physics:
A Comprehensive Study Based on Physics Abstracts. Am. Inst.
Physics. 1964. (AIP/DRP PA1 1964)
   See. Wa p. 72 (note).

CURRENT INFORMATION; SELECTION AIDS

   See: AIP Checklist of Books for an Undergraduate Physics Library.
Am. Inst. Physics, 1963.

Current issues of:

American Journal of Physics. v. 1, 1933-
Comments on Nuclear and Particle Physics. v. 1, 1967-
Comments on Solid State Physics. v. 1, 1968-
Contemporary Physics. v. 1, 1959-
Journal of Physics. v. 1, 1968-
Physical Review. v. 1, 1893-   .
Physics Today. v. 1, 1948-   .
Reviews of Modern Physics, v. 1, 1929-
Review of Scientific Instruments. v. 1, 1930-

SECTION D. CHEMISTRY

Chemistry is the science which deals with the constitution of matter
and the changes which it undergoes under various influences. According to the Westheimer Report (Chemical & Engineering News 43:86,
(Nov. 29) 1965) the changing nature of chemistry is reflected in a
subdivision into such areas as: structural chemistry; synthetic chemistry;
chemical dynamics; liquid, solid and surface chemistry; theoretical
chemistry; and nuclear chemistry.   The classical divisions of chemistry
include:
  Inorganic chemistry: study of the nature, properties and reactions
      of materials and their compounds.
  Organic chemistry: study of the nature, properties, and reactions
      of complex substances found in nature, chiefly compounds of carbon.
  Analytical chemistry: determination of the composition of compounds.
  Physical chemistry: study of the theoretical aspects of chemistry
      using fundamental laws formulated by physicists.
  Biochemistry: chemistry of living things.
  Radiochemistry: deals with the chemistry of the effects of high-
      energy radiation on matter.

GUIDES TO THE LITERATURE

D1. Bottle, R. T. ed. Use of Chemical Literature. 2d ed. Archon,
1968. 016.54 B659u
  See: Wi-ED2. Wa p. 86.

D2. Burman, C. R. How to Find Out in Chemistry; a Guide to Sources
of Information. Pergamon, 1965. 016.54 B92h
  See: J. Chem. Doc. 5:192, 1965; Chem. & Eng. News 43:66-7,
1965. Wa p. 87.

D3. Cahn, R. S. Survey of Chemical Publications and Report to the
Chemical Society. London, Chemical Society. 1965.
  See: Sp. Libs. 57:126, 1966. Wi-1ED1.

D4. Crane, E. J. et al. A Guide to the Literature of Chemistry.
2d ed. Wiley, 1957. 016.54 C85g
  See: Science 126:127, 1957. Wi-ED3. Wa p. 88.

D5. Mellon. M.G. Chemical Publications; Their Nature and Use. 4th ed. McGraw-Hill, 1965. 016.54 M48c
See: J. Chem. Ed. 43:A544, 1966. Wi-ED4. Wa p. 89.

D6. Van Luik, J. et al. Searching the Chemical and Chemical Engineering Literature. 2d ed. Purdue Univ., 1957. 016.5405 V32s
See: Wa p. 90.

D7. American Chemical Society. Literature of Chemical Technology. Wash., 1969. (Advances in Chemistry Series No. 78)

D8. American Chemical Society. Searching the Chemical Literature. Wash., 1961. (Advances in Chemistry Series No. 30) A540 Am35
See: Lib. J. 86:3932-3, 1961. Wi-ED1. Wa p. 85.

D9. American Chemical Society. Literature Resources for Chemical Process Industries. Wash., 1954 (Advances in Chemistry Series No. 10) 660 Am3121
See: Wa p. 350.

D10. American Chemical Society. A Key to Pharmaceutical and Medicinal Chemistry. Wash., 1957. (Advances in Chemistry Series No. 16) 615.04 Am3k
See: Wa p. 193.

BIBLIOGRAPHIES, INDEXES, ABSTRACTS

General

D11. American Chemical Society. Directory of Graduate Research. Wash., v.1, 1953- . (Biennial. Title varies) 016.54 Am36f
See: Chem. & Eng. News 46:65, (Dec. 4) 1967. Wi-ED11.

D12. Bolton, H.C. Select Bibliography of Chemistry 1492-1902. Wash., Smithsonian Inst., 1893-1904. 4v. 506 Sm6m
See: Wi-ED5. Wa p. 86.

D13. British Abstracts. London, 1926-53. (title varies) 540.5 BR
Sections include: A.I: General, physical and inorganic chemistry; A.II: Organic chemistry; A.III: Physiology, biochemistry, anatomy; B.I: Chemical engineering and industrial inorganic chemistry, including metallurgy; B.II: Industrial organic chemistry; B.III: Agriculture, foods and sanitation; C.: Analysis and apparatus.
See: Wi-ED16. Wa p. 86.

D14. Chemical Abstracts. Easton, Pa. American Chemical Society,
v.1, 1907-   . (Weekly on a biweekly cycle) Available in print, on
microfilm, and in five individual section groupings.   540.6 AMC
  See: Information Services from Chemical Abstracts Service. Am.
Chem. Soc., 1969. (revised annually) Wi-ED17. Wa p. 87.

D15. Chemical Abstracts. List of Periodicals with Key to Library
Files. 1961. Wash., Am. Chem. Soc., 1962. Supplements, 1962-   .
(A reprint of part of the annual author and patent index volume of
Chemical Abstracts) 540.6 AMC1
  See: Wi-ED12. Wa p. 87.

D15a. Access. Wash., Am. Chem. Soc., v.1, 1969-   .)Planned as
a replacement of quinquennial list of periodicals described above.
(No. D15)
  See: Chem. & Eng. News 47:46, (Mar. 31) 1969.

D16. Chemical Titles. Am. Chem. Soc., v.1, 1960-   . (Available
in print and on magnetic tape) 540.5 CHT
  See: Wi-ED18. Wa p. 87.

D17. Chemisches Zentralblatt. Berlin, Verlag Chemie. v.1, 1830-   .
540.5
  See: Wi-ED19. Wa p. 87.

D18. Current Chemical Papers. London, Chemical Society. v.1,
1954-   . 540.5 CU
  See: Wi-ED16 (note). Wa p. 88.

D19. Current Contents: Chemical Sciences. Institute for Scientific
Information. v1, 1958-   . (Title varies) 505 CUR
  See: Wa p. 88, 142.

D20. Journal of Applied Chemistry. London, Society of Chemical
Industry. v. 1, 1954-   . (Abstract section continues British Abstracts,
Section B I-II, 1871-1953) 660.5 JOUA
  See: Wa p. 85.

D21. Patent Index to Chemical Abstracts, 1907-1936. Edwards, 1944.
540.6 AMCindex
  See: Wi-ED17.

D22. Collective Numerical Patent Index to Chemical Abstracts,
v.31-40. 1937-1940. Am. Chem. Soc., 1949. 540.6 AMCindex
  See: Wi-ED17.

See also:

    Bulletin Signalétique. 7-8: Chimie (No. A18)
    International Catalogue of Scientific Literature. D:Chemistry.
      (No. A16)
    Referativnyi Zhurnal. Khimiya. (No. A19)

Special

D23. Chemical-Biological Activities. Am. Chem. Soc. v.1, 1965-
574.19205 CHE
  See: J. Chem. Doc. 3:81-5, 1963. Chem. & Eng. News 42:64-5,
(Nov. 16) 1964. Wi-EC12.

D24. East European Science Abstracts. London, Translation and
Technical Information Services. v.1, 1965- . 660.5 EA
  See: Sp. Libs. 56:410, 1965.

D25. Howell, M.G. et al. eds. Formula Index to NMR Literature Data.
Plenum, 1965- . 547 H83f

D26. Index Chemicus. Institute for Scientific Information. v.1,
1960- . (Cumulated annually as Encyclopaedia Chimica Inter-
nationalis). 540.5 INC
  See: J. Chem. Doc. 8:74-80, (May) 1968. Wi-ED21. Wa p. 89.

D27. Index to Reviews, Symposia Volumes and Monographs in Organic
Chemistry. Pergamon, 1962. 016.547 K52i
  See: Choice 4:804, 1967. Wi-ED71.

D28. Passwater, R.A. Guide to Fluorescence Literature. Plenum,
1967. 016.53535 P26g
  See: New Tech. Bks. 52:265, 1967.

D29. Polymer Science & Technology (POST). Am. Chem. Soc.
v.1, 1967- . Issued in two sections: POST-J and POST-P.
Available in printed form and on magnetic tape.

D30. Purdue University. Thermophysical Properties Research
Center. Thermophysical Properties Research Literature Retrieval
Guide. 2d ed. Plenum, 1967- 3v. projected. 016.6201129 P97r

D31. Signeur, A.V. Guide to Gas Chromatography Literature.
Plenum, 1964- 016.543 Si2g
  See: New Tech. Bks. 50:50, 1965. Wa p. 101 (note).

See listing of annual reviews in:
    Crane, E. J. et al. A Guide to the Literature of Chemistry.
       p. 207-212. (No. D4)
    Mellon, M. G. Chemical Publications. 3d ed. p. 111-113.
    (No. D5)

D32. Accounts of Chemical Research. Wash., Am. Chem. Soc.,
v. 1, 1968- . 540.7205 ACC

D33. Advances in Analytical Chemistry and Interscience, v. 1,
1960- . 543 Ad95
  See: Wa p. 100.

D34. Advances in High Temperature Chemistry. Academic, v. 1,
1967- . 541.36 Ad9

D35. Advances in Inorganic Chemistry and Radiochemistry. Academic,
v. 1, 1959- . 540 Ad9
  See: Wa p. 102 (note).

D36. Advances in Macromolecular Chemistry. Academic, v. 1, 1968-

D37. Advances in Organic Chemistry. Interscience, v. 1, 1960- .
547 Ad95
  See: Wa p. 104.

D38. Advances in Physical Organic Chemistry. Academic, v. 1,
1963- . 547.1 Ad9
  See: Wa p. 104.

D39. Advances in Polymer Science. Springer-Verlag, v. 1, 1958-
541.7 F77

D40. Advances in Quantum Chemistry. Academic, v. 1, 1964-
541.383 Ad9

D41. Annual Reports on the Progress in Chemistry. London,
Chemical Society, v. 1, 1905- . 540.6 CH
  See: Wa p. 94.

D42. Annual Review of Biochemistry. Annual Reviews, v. 1, 1932-
612.01 An78
  See: Wa p. 149.

D43. Annual Review of Physical Chemistry. Annual Reviews, v. 1, 1950- . 541 An78
  See: Wa p. 97.

D44. Chemical Reviews. Wash., Am. Chem. Soc., v.1, 1924- .
540.5 CHER

D45. National Research Council. Committee for the Survey of Chemistry. Chemistry: Opportunities and Needs; a Report on Basic Research in the U.S. Chemistry. Wash., 1965. (NRC Pub. 1292) 540.72 N21c

D46. Progress in Physical Organic Chemistry. Interscience, v.1, 1963- . 547.1 P94
  See: Wa p. 104 (note).

D47. Progress in Solid State Chemistry. Pergamon, v. 1, 1964-
539.1 P94.

D48. Survey of Progress in Chemistry. Academic, v.1, 1963-
  See: Wa p. 94.

HISTORIES

D49. Bäumler, E. A Century of Chemistry. Dusseldorf, Econ, 1968. (Translated from the 1963 German edition)
  See: Science 163:558, 1969.

D50. Chymia; Annual Studies in the History of Chemistry. Univ. of Penn. Pr. v.1, 1948- . 540.9 C479
  See: Wa p. 95.

D51. Haynes, W. American Chemical Industry. Van Nostrand, 1945-54. 6v. 660 H33a
  See: Chem. & Eng. News 23:1590-2, (June 10) 1945. Wi-ED47.

F52. Knight, D.M. ed. Classical Scientific Papers - Chemistry. Am. Elsevier. 1968.
  See: Lib. J. 93:3794, 1968; Science 162:110, 1968.

D53. Kopp, H. Geschichte der Chemie. Braunschweig, 1843-47. 4v. 540.9 K83g
  See: Chem. & Eng. News 43:47, (Jan. 4) 1965.

D54. Leicester, H.M. & Klickstein, H.S. Source Book in Chemistry, 1400-1900. McGraw-Hill, 1952. 540.9 L53s
  See: Wa p. 96.

D55. Leicester, H.M. ed. Source Book in Chemistry, 1900-1950. Harvard Univ. Pr., 1968. 540.8 L53so
  See: Science 162:110-111, 1968; Lib. J. 93:2673-4, 1968.

D56. Moore, F.J. History of Chemistry. 3d ed. McGraw-Hill, 1939. 540.9 M78h3

D57. Multhauf, R.P. The Origins of Chemistry. Watts, 1967. 540.9 M91o
  See: Science 158:364, 1967.

D58. Partington, J.R. History of Chemistry. St. Martins, 1962- 4v. projected. 540.9 P25h
  See: Chem. & Eng. News 43:47, (Jan. 4) 1965; Science 139:1192-3, 1963; J. Chem. Ed. 42:346, 1965. Wi-ED83. Wa p. 96.

D59. Pennsylvania. University. Edgar Fahs Smith Memorial Library. Catalog of the Edgar Fahs Smith Memorial Collection in the History of Chemistry. G.K. Hall, 1960. q016.5409 P38c
  See: Wa p. 96.

D60. Weeks, M.E. Discovery of the Elements. 7th ed. Easton, Pa. Journal of Chemical Education, 1968. 546 W41d
  See: Science 162:110-111, 1968.

  Consult the bibliographical aids for subject bibliographies and journals that review special areas.

DICTIONARIES & ENCYCLOPEDIAS

  General

D61. Advances in Chemistry Series; a Continuing Series of Books Published by the American Chemical Society. Wash., 1950- . (Early titles available in microfiche only)

D62. Bennett, H., ed. Concise Chemical and Technical Dictionary. 2d ed., Chemical Pub., 1962. 540.3 B43c
  See: Sub. Bks. Bull. 59:834, 1963. Wa p. 90.

D63. Chemical Technology: An Encyclopedia Treatment. The Economic Application of Modern Technological Developments. Barnes & Noble, 1968- . (8v. projected)

D64. Condensed Chemical Dictionary. 7th ed. Reinhold, 1966. 540.3 C751
See: New Tech. Bks. 52:21, 1967; Lib. J. 93:1585, 1968. Wi-1ED4

D65. Encyclopedia of Chemical Technology (Kirk-Othmer Encyclopedia). 2d ed. Interscience, 1963- . 660 En19
See: Wi-1EI9

D66. Encyclopedia of Chemistry. (G.C. Clark, ed.) 2d ed. Reinhold, 1966. 540.3 En19
See: Wi-1ED6. For review of 1st ed. see: Sub. Bks. Bull. 53:543, 1957.

D67. Grant, J. Hackh's Chemical Dictionary. 4th ed. McGraw-Hill, 1968. 540.3 H11
See: Lib. J. 93:4171, 1968.

D68. Hampel, C.A. ed. Encyclopedia of Chemical Elements. Reinhold, 1968. 546.03 H18e
See: Lib. J. 93:4540, 1968.

D69. Kingzett, C.T. Kingzett's Chemical Encyclopedia: A Digest of Chemistry & Its Industrial Applications. 9th ed. London, Baillière, 1966. 540.3 K61c
See: Lib. J. 93:1585, 1968.

D70. Miall, L.M ed. A New Dictionary of Chemistry. 4th ed. Wiley, 1968. 540.3 M58n
See: Choice 5:1430, 1969; Lib. J. 93:4540, 1968.

D71. Van Nostrand's International Encyclopedia of Chemical Science. Van Nostrand, 1964. (Includes multilingual indexes: French, German, Russian, Spanish) 540.3 In8
See: New Tech Bks. 49:160, 1964; Am. Scientist 52:457A-8A, 1964. Wi-ED23.

D72. White, J.H. A Reference Book of Chemistry. Philosophical, 1965. 540.3 W58r
See: RQ 7:88-9, 1967; Choice 4:410, 1967.

Synonyms & Trade names

D73. Gardner, W. Chemical Synonyms and Trade Names. 5th ed.
Van Nostrand, 1948. 660.3 G17c
  See: Wi-ED33. Wa p. 92.

D74. Haynes, W. Chemical Trade Names and Commercial Synonyms.
2d ed. Van Nostrand, 1955. 660.3 H33c
  See: Wi-ED47. Wa p. 353.

D75. Synthetic Organic Chemical Manufacturers Assn. SOCMA Handbook: Commercial Organic Chemical Names. Wash., Amer. Chem.
Soc., 1965. 547.014 Sy7s
  See: Wi-1ED18.

D76. Zimmerman, O. T. & Lavine, I. Handbook of Material Trade
Names. 2d ed. Dover, N.H., Industrial Research Service, 1953.
Supplements, 1956-   608 Zi6h
  See: Wi-CH149. Wa p. 177.

Foreign language

D77. Chu, C. ed. Applied English-Chinese & Chinese-English
Chemical Dictionary. Thousand Oaks, Calif., Oriental Publication
Service. 1965.

D78. Yang, P. et al. English-Chinese Dictionary of Chemistry and
Chemical Engineering. Thousand Oaks, Calif., Oriental Publication
Service, 1965. 540.3 C36e

D79. Patterson, A.M. A French-English Dictionary for Chemists.
2d ed. Wiley, 1954. 540.3 P27f
  See: Wi-ED37. Wa p. 93.

D80. Ernst, R. ed. Dictionary of Chemistry... V.1: German-
English. V.2: English-German. Wiesbaden, Brandstetter, 1961-63.
2v.
  See: Wi-ED38. Wa p. 92.

D81. Neville, H.H. et al. A New German-English Dictionary for
Chemists. Van Nostrand, 1964. 540.3 N41n
  See: Wa p. 93.

D82. Patterson, A.M. German-English Dictionary for Chemists.
3d ed. Wiley, 1950. 540.3 P27g
  See: Wi-ED39. Wa p. 93.

D83. Rompp, H. Chemie Lexikon. 6th ed. Stuttgart, Franckh'sche Verlag, 1966. 4v. (English-German/German-English) 540.3 R66c

D84. Wohlauer, G. & Gholston, H.D. eds. German Chemical Abbreviations. Sp. Lib. Assn., 1965. 540.148 W82g

D85. Callaham, L.I. Russian-English Chemical and Polytechnical Dictionary. 2d ed. Wiley, 1962. 540.3 C13r
See: Science 140:654, 1963. Wi-ED44. Wa p. 93.

D86. Hoseh, M. & Hoseh, M.L. Russian-English Dictionary of Chemistry and Chemical Technology. Reinhold, 1964. 540.3 H79r
See: Chem. & Eng. News 42:79, (Dec. 7) 1964. Wa p. 94.

D87. Karpovich, E.A. & Karpovich, V.V. Russian-English Chemical Dictionary. 2d ed. Technical Dictionaries, 1963. (alternate entry: Carpovich, E.A. & Carpovich, V.V.) 540.3 K14r
See: Science 136:515-6, 1962. Wi-ED44. Wa p. 93.

D88. Goldberg, M. Spanish-English Chemical and Medical Dictionary. McGraw-Hill, 1952. English-Spanish Chemical and Medical Dictionary McGraw-Hill, 1947. 610.3 G56s
See: New Tech. Bks. 37:85, 1952. Wa p. 185.

D89. Dictionary of Chemistry and Chemical Technology, in Six Languages: English, German, Spanish, French, Polish, Russian. Pergamon, 1966. 540.3 D56
See: New Tech. Bks. 51:236. 1966.

D90. Elsevier's Dictionary of Industrial Chemistry in Six Languages: English/American, French, Spanish, Italian, Dutch, and German. Elsevier, 1964. 2v.
See: Wa p. 352.

D91. Fouchier, J. & Billet, F. Chemical Dictionary. 2d ed. Amsterdam, Netherlands Univ. Pr., 1961. (English-German-French) 540.3 F82d
See: Science 136:520, 1962. Wi-ED40. Wa p. 92.

D92. Mayer, A.W. Chemical-Technical Dictionary. Chemical Pub., 1942. (German-English-French-Russian) 540.3 M45ch
See: New Tech. Bks. 28:7, 1943. Wi-ED42.

D93. Russian-Chinese-English Chemical and Technical Dictionary. London, Scientific Information Consultants, 1965. 540.3 R92
See: New Tech. Bks. 51:126, 1966.

COMPREHENSIVE WORKS

Analytical chemistry

D94. Analytical Chemistry, U.S.S.R. Academy of Sciences of the
U.S.S.R. 1960- . (Approximately 50 monographs projected. Catalogued
and classified as separates)
  See: Science 148:1452-3, 1965.

D95. Encyclopedia of Industrial Chemical Analysis. Interscience.
v.1, 1966- . (15v. projected) 543.03 En19
  See: Science 154:640-1, 1966; Sp. Libs. 57:670, 1966.

D96. Jolly, S.C. comp. Official, Standardized and Recommended
Methods of Analysis. Heffer, 1963. 543 Jo13o
  See: Wa p. 100 (note).

D97. Kolthoff, I.M. & Elving, P.J. Treatise on Analytical Chemistry.
Interscience, 1959-   543 K83t
  See: Wa p. 100.

D98. Standard Methods of Chemical Analysis. 6th ed. Van Nostrand,
1962- . 543 W69c
  See: Chem. & Eng. News 40:112-3, (Sept. 10) 1962. Wa p. 100.

D99. Wilson, C.L. ed. Comprehensive Analytical Chemistry.
Van Nostrand, 1959- . 543 W69c
  See: New Tech. Bks. 46:111-2, 1961. Wa p. 100.

Biochemistry

    See also: BIOLOGICAL SCIENCES, AGRICULTURAL SCIENCES,
and MEDICAL SCIENCES

D100. Florkin, M. & Mason, H.S. eds. Comparative Biochemistry.
Academic, 1960-1964. 7v. 574.192 F66c
  See: Science 137:745, 1962. Wa p. 148.

D101. Florkin, M. & Stotz, E.H. eds. Comprehensive Biochemistry.
Elsevier, 1962- . 547.1 F66c
  See: Sci. Am. 220:126-7, (Feb.) 1969; Science 140:1201-3, 1963.
Wa p. 149 (note).

D102. Methods of Biochemical Analysis. Interscience, v.1, 1954-
612.01 M56

D103. National Research Council. Specifications and Criteria for Biochemical Compounds. 2d ed. Wash., 1967. (NAS-NRC Pub. No. 1344) 574.192 N21s

D104. Williams, R.J. & Lansford, E.M., Jr., eds. Encyclopedia of Biochemistry. Reinhold, 1967.
  See: Bioscience 18:58, 1968; New Tech. Bks. 52:201, 1967.

Inorganic chemistry

D105. Gmelin, L. Gmelin's Handbuch der anorganischen Chemie. 8th ed. Berlin, Verlag, Chemie. v.1, 1924- . Supplements, 1937- . 546 G52h8
  See: Sp. Libs. 50:492-6, 1959. Wi-ED61. Wa p. 101-2.

D106. Jacobson, C.A. comp. Encyclopedia of Chemical Reactions. Reinhold, 1946-59. 8v. 546 J15e
  See: Wa p. 97.

D107. Jonassen, H.B. & Weissberger, A. eds. Technique of Inorganic Chemistry. Interscience. v.1, 1963- . 546 T22
  See: Chem. & Eng. News 42:60, (Nov. 2) 1964.

D108. Mellor, J.W. A Comprehensive Treatise on Inorganic and Theoretical Chemistry. Longmans, 1922-37. 16v. Supplements, 1956-546 M48c
  See: Wi-ED62. Wa p. 102.

D109. Pascal, P. Nouveau Traité de Chimie Minérale. Paris, Masson, 1956- . 546 P26t
  See: Wa p. 108.

D110. Sneed, M.C. et al. Comprehensive Inorganic Chemistry. Van Nostrand, 1955- . 546 Sn2c
  See: Wa p. 102.

Organic Chemistry

D111. Beilstein, F. Beilstein's Handbuch der organischen Chemie. 4th ed. Berlin, Springer, 1918-1940. 31v. Supplement I, 1928-38. 27v. Supplement II, 1941-57. 29v. Supplement III, 1958- . 547 B393h
  See: Huntress, E.H. A Brief Introduction to the Use of Beilstein's Handbuch ... Wiley, 1938. 547 B393h4Yh2

Deutsche Chemische Gesellschaft. System der organischen
Verbindungen. Berlin, Springer, 1929.
  See also: Wi-ED64. Wa p. 103.

D112. Elsevier's Encyclopedia of Organic Chemistry. Elsevier.
v. 12-14, 1940- . 547 E176
  See: Wi-ED69. Wa p. 103.

D113. Dictionary of Organic Compounds; the Constitution and Physical,
Chemical and other Properties of the Principal Carbon Compounds
and their Derivatives, Together with the Relevant Literature references.
4th ed. Oxford Univ. Pr., 1965. 5v. Supplement, 1965- . (annual)
547.03 H36d
  See: Science 150:1280, 1965. Wi-ED68.

D114. Encyclopedia of Polymer Science and Technology. Interscience.
v. 1, 1964- . (10v. projected over 5 year period) 547.84 En19
  See: Science 147:386-7, 1965. Wi-EI63. Wa p. 402.

D115. Fox, D. ed. Physics and Chemistry of the Organic Solid State.
Interscience, 1963- . 548.8 F83p.

D116. Grignard, V. et al. Traité de Chimie organique. Paris, Masson.
1935-54. 23v. in 27. 547 G87t
  See: Wa p. 103.

D117. Houben-Weyl's Methoden der organischen Chemie. 4th ed.
Stuttgart, Thieme, 1952- . (16v. in numerous parts projected)
547 W54m
  See: Wi-ED73. Wa p. 103.

D118. Patterson, A.M. et al. The Ring Index: A List of Ring Systems
Used in Organic Chemistry. 2d ed. Wash., Amer. Chem. Soc., 1960.
Supplements, 1963- . 547 P27r
  See: Chem & Eng. News 38:62, (Aug. 1) 1960. Wa p. 105.

D119. Rodd's Chemistry of Carbon Compounds; a Modern Comprehensive Treatise. 2d ed. Elsevier, 1964- . 547 R61c
  See: New Tech. Bks. 51:125, 1966. Wa p. 105.

D120. Theilheimer, W. Synthetic Methods of Organic Chemistry.
Interscience. v. 1, 1946- . (annual since 1951) 547 T34sEw
  See: Wa p. 105.

D121. UV Atlas of Organic Compounds. Plenum. V. 1, 1966-
q535.844 Uv1
  See: Am. Scientist 56:289A, 1968.

D122. Weissberger, A. ed. Technique of Organic Chemistry. Interscience, 1949- . (some volumes are in 2d ed. and others in 3d ed.) 547 W43t
  See: Wa p. 104.

D123. Gowen, J.E. & Wheeler, T.S. Name Index of Organic Reac-2d ed. Longmans, 1960. 547.2 W56n
  See: Wa p. 104-5.

D124. Krauch, H. & Kunz, W. Organic Name Reactions. Wiley, 1964. 547.139 K86n.
  See: Chem. & Eng. News 43:59, (June 7) 1965. Wi-ED72.

D125. Surrey, A.R. Name Reactions in Organic Chemistry. 2d ed. Academic, 1961. 547 Su7n
  See: J. Chem. Ed. 39:326, 1962.

D126. Organic Reactions. Wiley, v.1, 1942- . (annual) 547 Or32
  See: Wa p. 105.

D127. Organic Syntheses. Wiley. v.1, 1921- . (annual) Collective volumes. 1932- . (published after every tenth annual volume) 547 Or31
  See: Wa p. 105.

Physical chemistry

    See: International Encyclopedia of Physical Chemistry and
       Chemical Physics (No. C56)

D128. Hampel, C.A. ed. Encyclopedia of Electrochemistry. Reinhold. 1964. 541.3703 H18e
  See: Chem. & Eng. News 43:82, (Sept. 13) 1965; New Tech. Bks. 49:306-7, 1964. Wi-EI57. Wa p. 98.

D129. Handbook of Thermophysical Properties of Solid Materials. Pergamon, 1961. 620.1129 Ar5h

D130. Porter, M.W. & Spiller, R.C. The Barker Index of Crystals: a Method for the Identification of Crystalline Substances. Heffer, 1951-64. 3v. in 7. 548.7 P83b
  See: Science 149:45-6, 1965. Wa p. 107.

D131. Purdue University. Thermophysical Properties Research Center. Thermophysical Properties of High Temperature Solid Materials. Macmillan, 1967- . 620.1129 P97t

HANDBOOKS & TABLES

See also:
Handbook of Chemistry and Physics. (No. A79)
The handbooks listed in the sections on GENERAL SCIENCE, PHYSICS, AGRICULTURAL SCIENCES, and MEDICAL SCIENCES

D132. Handbook of Chemistry. Comp. by N.A. Lange et al. McGraw-Hill, 1st ed., 1934- . (rev. 10th ed., 1967) 660 H191
See: Lib. J. 92:2752, 1967; New Tech. Bks. 52:231, 1967. Wi-ED49. Wa p. 95.

D133. CRC Handbook of Biochemistry with Selected Data for Molecular Biology. Chemical Rubber Co., 1968.
See: Research/Development 19:80, (Aug.) 1968.

D134. CRC Handbook of Laboratory Safety. Chemical Rubber Co., 1967. 542.1 St3c
See: Science 159:1451-2, 1968.

D135. Handbook of Biochemistry and Biophysics. N.Y., World Pub., 1966. 574.19 D18h
Companion volume: Methods and References in Biochemistry and Biophysics. 1966. (No. G76)
See: Choice 4:147-8, 1967.

D136. International Tables for X-ray Crystallography. International Union of Crystallography, 1952-1962. 3v. 548.7 In82
See: Science 118:222, 1953. Wi-EE70. Wa p. 107.

D137. Keller, Roy A. Basic Tables in Chemistry. McGraw-Hill, 1967. 540.83 K28b
See: Choice 5:34, 1968.

D138. Linke, W.F. Solubilities, Inorganic and Metal-organic Compounds; a Compilation of Solubility Data from the Periodical Literature. 4th ed. Van Nostrand, 1958-66. 2v. 541.8 Sets
A continuation of: Seidell, A. Solubilities of Inorganic and Metal Organic Compounds. 3d ed. 1940-1. 2v.
See: Wi-ED54. Wa p. 98.

D139. Meites, L. ed. Handbook of Analytical Chemistry. **McGraw-Hill.**
1963. 543 M479h
  See: Am. Scientist 51:318A, (Sept) 1963. Wi-ED51. Wa p. **100.**

D140. Merck Index: An Encyclopedia of Chemicals and Drugs. **8th ed.**
Rahway, N.J., Merck. 1968. 615.1 M537
  See: Choice 5:1430, 1969.

D141. Polymer Handbook. Wiley, 1966. 547.8402 B73p
  See: Science 153:1372, 1966. New Tech. Bks. 51:157-8, 1966.
Wi-1EI12.

D142. Rosin, J. Reagent Chemicals and Standards; with Methods of
Testing and Assaying them   5th ed. Van Nostrand, 1967. 543
R731r

D143. Samsonov, G.V. ed. Handbook of the Physicochemical Properties of the Elements. Plenum, 1968.
  See: New Tech. Bks. 53:126, 1968; Physics Today 21:97, (Sept.) 1968.

D144. Sax, N.I. ed. Dangerous Properties of Industrial Materials.
3d ed. Reinhold, 1968. 331.823 Sa98h.

D145. Snell, F.D. & Snell, C.T. Dictionary of Commercial Chemicals. 3d ed. Van Nostrand, 1962. 661 Sn2c
  See: Chem. & Eng. News 40:88, (Oct. 29,) 1962. Wi-ED55. Wa p. 357.

D146. Stephen, H. & Stephen, T. eds. Solubilities of Inorganic and
Organic Compounds. Pergamon, 1963-   547.134 M85sE
  See: Wi-ED56. Wa p. 98.

D147. Szymanski, H.A. Infrared Band Handbook. Plenum, 1963.
Supplements, 1964- . 535.842 Sz93i
  See: Wi-Ed57. Wa p. 80.

D148. Szymanski, H.A. & Yelin, R.E. NMR Band Handbook.
Plenum, 1968. 538.3 Sz9n
  See: New Tech. Bks. 53:127, 1968.

D149. Yukawa, Y. ed. Handbook of Organic Structural Analysis.
N.Y., Benjamin. 1965. 547.12 Y91h

INFORMATION ACTIVITIES

D150. Journal of Chemical Documentation. Easton, Pa., American Chemical Society, v. 1, 1961- . 540.5 JOCD
  Note in particular the issues which cover symposia on this subject. See: Wa p. 85.

D151. An Overview of Worldwide Chemical Information Facilities and Resources. Wash., Clearinghouse, 1968. (PB 176-160) 010.78 OV2.
  See: Sci. Info. Notes 9:24, (Oct./Nov.) 1967.

D152. U.S. National Bureau of Standards. File Organization for a Large Chemical Information System. Wash., 1966. (NBS Technical Note 285) 660 Un3t

NOMENCLATURE

D153. American Chemical society. Chemical Nomenclature. Wash., 1953. (Advances in Chemistry Series No. 8) 540 Ad95

D154. Cahn, R.S. Introduction to Chemical Nomenclature. 3d ed. Plenum, 1968. 540.14 C11i

D155. International Union of Pure and Applied Chemistry. Nomenclature of Inorganic Chemistry; Definitive Rules for Nomenclature of Inorganic Chemistry. London, Butterworths. 1959. 546.014 In8n
  See: Wi-ED63. Wa p. 101.

D156. International Union of Pure and Applied Chemistry. Nomenclature of Organic Chemistry. 2d ed. London, Butterworths. 1965-66. 3v. 547.014 In8n
  See: New Tech. Bks. 51:264, 1966.

D157. International Union of Pure and Applied Chemistry. Rules for I.U.P.A.C. Notation for Organic Compounds. Longmans, 1961. 547.0148 In8r
  See: Wa p. 103.

  See discussion of nomenclature in: Burman, C.R. How to Find Out in Chemistry. p. 99-100. (No. D2)

FORMULARIES

D158. Bennett, H. Chemical Formulary. Brooklyn, Chemical
Formulary Co., 1933- . (cumulative index for v. 1-10)
660 B43c
   See: Sub. Bks. Bull. 17:26, 1946. Wi-EA208. Wa p. 353.

D159. Henley's Twentieth Century Book of Formulas. rev. ed.
N.Y., Henley, 1947. 603 H62e
   See: Wi-EA209. Wa p. 173.

D160. Hopkins, A.A. ed. Standard American Encyclopedia of Formulas.
Grossett & Dunlap, 1953. 603 H77s
   See: Wa p. 173.

D161. Minrath, W.R. ed. Van Nostrand's Practical Formulary.
Van Nostrand, 1957. 603 M663v

INDUSTRIAL INFORMATION & DIRECTORIES

See also section on GENERAL SCIENCE

D162. American Chemical Society. Reagent Chemicals. Wash., 1941-
(Title varies) Fourth edition (1968) supplemented in Analytical Chemis-
try. 543 Am33a
   See: Chem. & Eng. News 47:76, (Oct. 14) 1968. Choice 5:1423, 1969.

D163. Chem Sources. Flemington, N.J., Directors Pub., 1958- .
q661 C424

D164. Chemical and Engineering News. v. 1, 1923- . Note special
issues containing "Facts & Figures for the Chemical Process Industries" and "Outlook" (view of coming year). 540.5 IN

D165. Chemical Engineering Catalog. Reinhold, 1916- . Supplement I: The Flow Sheet. 1931- . Supplement II: Process Engineering. 1946- . 660 C42

D166. Chemical Guide to Europe. Pearl River, N.Y., Noyes Development Co., 1963-
   See: Sci. Ref. Notes 10:9, (July/Oct.) 1963. Wi-ED60 (note).

D167. Chemical Guide to the United States. Pearl River, N.Y., Noyes
Development Co., 1962- . 660.58 C4215
   See: Sci. Ref. Notes 10:10, (July/Oct.) 1963. Wi-ED60.

D168. Chemical Industry Directory and Who's Who. London, Benn Bros., 1923-   . (title varies) q540 C423
  See: Wa p. 353.

D169. Chemical Markets Abstracts. Foster D. Snell, 1950-   .
  See: Wa p. 351.

D170. Chemical Materials Catalog and Directory of Producers. Reinhold, 1949/50-   . 660 C418

D171. Chemical Statistics Directory: ... an Index to Government Statistics of the Chemical Industry. 1945-   . (irregular) 338.4 Un3235c
  See: Chem. & Eng. News 40:30, (Feb. 26) 1962.

D172. Information on International Scientific Organizations, Services, and Programs for Chemists, Chemical Engineers and Physicists, 1966-   . (Supersedes information booklets formerly prepared separately by the American Chemical Society and the American Institute of Physics) 506 INF
  See: Chem. & Eng. News 44:72, (Oct. 3) 1966.

D173. International Chemistry Directory. W.A. Benjamin, 1969-   .

D174. Manufacturing Chemists' Association.   The Chemical Industry Fact Book. Wash., 1953-   . 338.4 M31ch

D175. Patents for Chemical Inventions. Amer. Chem. Soc., 1964. (Advances in Chemistry Series, No. 46) 540 Ad95 No. 46.
  See: Chem. & Eng. News 43:88, (Mar. 8) 1965.

D176. Uniterm Index to the United States Chemical Patents. Wash., 1950-   .
  See: J. Chem. Doc. 8:23-5, (Feb. 1) 1968.

BIOGRAPHIES

D177. Chemical Who's Who. N.Y., Lewis Historical Co., 1928-56. 925.4 W62
  See: Wi-ED77. Wa p. 354 (note).

D178. Farber, E. Great Chemists. Interscience, 1961. 925.4 F22g
  See: Chem. & Eng. News 40:501-2, 1962. Wi-ED78. Wa p. 96.

D179. Farber, E. Nobel Prize Winners in Chemistry. 2d ed.
Abelard-Shuman. 1963. 925.4 F22n
  See: Science 139:623, 1963. Wi-ED79. Wa p. 96.

D180. Nobel Foundation. Nobel Lectures in Chemistry, 1901-1962.
Am. Elsevier, 1966. 3v.
  See: Am. Scientist 55:68A-9A, (Mar.) 1967.

CURRENT INFORMATION; SELECTION AIDS

Consult:

Advances in Chemistry Series (monographs)

Advisory Council on College Chemistry. Guidelines and Recommended Title List for Undergraduate Chemistry Libraries.
Stanford, Calif., 1966. (revision in preparation) 016.54 Ad9g

American Chemical Society. Selected Titles in Chemistry: an
Annotated Bibliography of Inexpensive Books for the General
Reader. Wash., 1966.

Wood, J.L. A Comprehensive List of Periodicals for Chemistry
and Chemical engineering. Library Trends 16:398-409, 1968

Current issues of such periodicals as:

American Chemical Society. Journal, v.1, 1879- .
Chemical Abstracts, v.1, 1907- .
Chemical & Engineering News, v.1, 1923- .
Chemistry, v.1, 1927- .
Education in Chemistry, v.1, 1964- .
Industrial and Engineering Chemistry, v.1, 1909- .
Journal of Chemical Documentation, v.1, 1961- .
Journal of Chemical Education (September issue lists titles
included in annual book exhibit) v.1, 1924-

SECTION E. ASTRONOMY

=====================

Astronomy is the science that treats of celestial bodies, their identi-
fication, motion, magnitude, distances, physical constitution, and the
space between them. Major subdivisions of current interest include:
   Aeronomy: study of the properties of the upper atmosphere and
      the physical processes occurring in it.
   Astrometry: branch that determines the positions and magnitudes
      of heavenly bodies by measurement of angles and time.
   Astrophysics: treats of the appearance and physical constitution
      of celestial bodies, their spectra, color, brightness, tem-
      perature, etc.
   Celestial mechanics: study of the motion of bodies in space as
      they are influenced by gravitation attraction.
   Cosmology: study of the origin and relationships of the universe.
   Radio astronomy: deals with radio signals which are produced in
      outer space.
   Selenology: that branch that deals with the moon.

BIBLIOGRAPHIES, INDEXES, ABSTRACTS

E1. Astronomischer Jahresbericht. v.1, 1899-   . 520.5 ASJ
   See: Wi-EB8. Wa p. 62.

E2. Bibliography of Natural Radio Emission from Astronomical
Sources. Cornell Univ. Pr. v.1, 1962-   . 016.523 B47

E3. Bibliographie Mensuelle de l'Astronomie. v.1-11, 1934-44.
(Continued since 1948 as a section of Bulletin Signalétique)
520.5 B1

E4. Houzeau, J.C. & Lancaster, A. Bibliographie générale de
l'Astronomie. Brussels, Hayez, 1882-1889. 2v. 016.52 H81
   See: Wi-EB4. Wa p. 62.

E5. Collard, A. L'Astronomie et les Astronomes. Brussels,
Van Oest, 1921. A520 C68q
   See: Wi-EB2.

See also:
  Bulletin Signalétique: Astronomie. Astrophysique. Physique du
      globe. (No. A18)
  Mathematical Reviews. (No. B10)
  Physics Abstracts. (No. C8)
  Referativnyi Zhurnal. Astronomiya. (No. A19)

REVIEWS & SURVEYS

E6. Advances in Astronomy and Astrophysics. Academic, v.1,
1962- . 520 Ad95
  See: Am. Scientist 51:92A-3A, 1963.

E7. Annual Review of Astronomy and Astrophysics. Annual Reviews,
v.1, 1963- . 523 An7
  See: Wa p. 63.

E8. Astrophysical Journal. Univ. of Chicago Pr., v.1, 1895-
520.5 AS

E9. Comments on Astrophysics and Space Physics. Gordon & Breach,
v.1, 1969-   . 523.0105 CO

E10. Vistas in Astronomy. Pergamon, v.1, 1955-   520 V82
  See: Wa p. 64.

E11. Yearbook of Astronomy. Norton, 1962-   523 Y23
  See: Science 140:288, 1963. Wa p. 64.

HISTORIES

E12. Abetti, G. The History of Astronomy. Schuman, 1952.
(translated from the Italian edition by B. B. Abetti) 520.9 Ab3s
  See: Wa p. 64.

E13. Berry, A. Short History of Astronomy. Macmillan, 1910.
520.9 D68c

E14. Cotter, C.H. A History of Nautical Astronomy. Am. Elsevier,
1968. 527.09 C82h
  See: Choice 5:1156, 1968.

71

E15. Vaucouleurs, G. de. Astronomical Photography; from the Daguerreo-type to the Electron Camera. (translated by R. Wright) Macmillan, 1961.
  See: New Tech. Bks. 47:154, 1962

E16. King, H.C. The History of the Telescope. Sky, 1955. 522.2 K58h

E17. Mueller, G. & Hartig, E. Geschichte und Literatur des Licht-wechsels. Leipzig, Poeschel, 1918-22. 3v. q523.84 M91g

E18. Prager, R. Geschichte und Literatur des Lichtwechsels der veranderlichen Sterne. 2d ed. Berlin, Dummler, 1934-57. 4v. q523.8 P88g
  A continuation of Mueller & Hartig's Geschichte.

E19. Pannekoek, A. A History of Astronomy. Interscience. 1961. (translated from the Dutch) 520.9 P19gE
  See: Wa p. 64.

E20. Shapley, H. & Howarth, H.E. Source Book in Astronomy. McGraw-Hill, 1929. 520 Sh2s
  See: Wi-EB21.

E21. Shapley, H. Source Book in Astronomy, 1900-1950. Harvard Univ. Pr., 1960. 520.8 Sh2s
  See: Science 133:696, 1961. Wi-EB22.

DICTIONARIES, ENCYCLOPEDIAS, DESCRIPTIVE WORKS

General

E22. American Institute of Physics. Glossary of Terms Frequently Used in Radio Astronomy. Wash., 1962.
  See: Wa p. 65.

E23. Dictionary of Astronomical Terms. Natural History Pr., 1966.
  See: Sky & Telescope 31:297, 1966. Wi-1EB3.

E24. Ernst, B. & de Vries, T.E. Atlas of the Universe. Thomas Nelson, 1961. 523.1 Er6wEw
  See: Sky & Telescope 23:341, 1962. Wi-EB23. Wa p. 65.

E25. Fairbridge, R.W., ed. The Encyclopedia of Atmospheric Sciences and Astrogeology. Reinhold, 1967. (Encyclopedia of Earth Sciences Series, v.2) 551.503 F15e
  See: Choice 4:100, 1968.

E26. The Flammarion Book of Astronomy. Simon & Schuster, 1964. q523 F61aEp
  See: Science 146: 1153, 1964. Wi-EB18. Wa p. 62.

E27. Franklin, K.L. Space Age Astronomy. Natural History Pr., 1964. 520 F85s

E28. Glasstone, S. Sourcebook of the Space Sciences. Van Nostrand, 1965. 523 G46s
  See: Science 149:1363, 1965; New Tech. Bks. 50:314-5, 1965.

E29. King, H.C. Pictorial Guide to the Stars. Crowell, 1967. 523.8 K58
  See: Sky & Telescope 35:176-7, (Mar.) 1968; New Tech Bks. 52:330, 1967.

E30. Larousse Encyclopedia of Astronomy. Prometheus, 1959. q520 R83aEg
  See: Subs. Bks. Bull. 56:410, 1960; Science 130:1704, 1959. Wi-EB13. Wa p. 63.

E31. Sky and Telescope. Telescopes; How to Make Them and Use Them. Macmillan, 1966. 522.2 Skyt
  See: Booklist 61:605-6, 1967.

E32. Solar System Radio Astronomy. Plenum, 1965. 523.016 Ad9s 523.016 Ad9s
  See: Science 151: 1376, 1966.

E33. Weigert, A. & Zimmermann, H. A Concise Encyclopedia of Astronomy. Am. Elsevier, 1968. (Translated from the second German edition of ABC der Astronomie; British edition has the title ABC of Astronomy)
  See: Booklist 65:566, 1969; Sky & Telescope 36:255, 1968; 35:314, 1968.

  Foreign language

E34. Chiu, H.Y. Chinese-English, English-Chinese Astronomical Dictionary. Plenum, 1966.
  See: Wi-1EB2.

E35. Hyman, C.J. Astronautics Dictionary: German-English, English-German. Plenum, 1968.

E36. Kramer, A.A. Russian-English Dictionary of Astronomy. Trenton, 1962. 520.3 K86r

E37. Kleczek, J. Astronomical Dictionary. In Six Languages: English, Russian, German, French, Italian, Czech. Academic, 1962. 520.31 K679
   See: Science 136:519-20, 1962. Wi-EB15. Wa p. 63.

COMPREHENSIVE WORKS

E38. Bowditch, N. American Practical Navigator. Wash., G.P.O., 1958.
   See: Wi-EB26.

E39. Handbuch der Astorphysik. Berlin, Springer, 1928-1936. 7v. in 10. 521 H191
   See: Wi-EB12.

E40. Kuiper, G.P. & Middlehurst, B.M. eds. The Solar System. Univ. of Chicago Pr., 1953- . 523.2 K958
   See: Science 119:548, 1954. Wa p. 66.

E41. Kuiper, G.P. & Middlehurst, B.M. Stars and Stellar Systems. Univ. of Chicago Pr., 1960- . (9v. projected)

HANDBOOKS; TABLES

E42. Allen, C.W. Astrophysical Quantities. 2d ed. Univ. of London Pr., 1963. 523.01 A14a
   See: Wa p. 65.

E43. Allen, R.H. Star-names and Their Meanings. Stechert, 1899. (Reprinted as: Star Names; Their Lore and Meaning. Dover, 1963) 523.89 A15s

E44. Cherrington, E.H., Jr. Exploring the Moon Through Binoculars. McGraw-Hill, 1969.
   See: Sky & Telescope 37:179, 1969.

E45. Moore, P., ed. A Handbook of Practical Amateur Astronomy. Norton, 1964. 523 M78h
   See: Am. Scientist 53:251A, 1965; Lib. J. 89:4922, 1964.

E46. Planetary, Lunar and Solar Positions. American Philosophical Society, 1962-64. (v.1: 601 BC to AD 1; v.2: AD2 to AD1649) 500 Am3m
  See: Physics Today 17:72, (June) 1964.

E47. Strand, K.A. ed. Basic Astronomical Data. Univ. of Chicago Pr., 1963. (v. 3 of the series Stars and Stellar Systems (No. E35)) 523 St78b
  See: Am. Scientist 52:204A, 1964.

E48. U.S. Nautical Almanac Office. Air Almanac. 1933, 1941-   . (British Air Almanac merged in 1953) 528.1 Un34am

E49. U.S. Nautical Almanac Office. American Ephemeris and Nautical Almanac. 1855- . Annual. (Merged with The Nautical Almanac (British) with the 1960 edition, but they will retain their individual identities.) 528.1 Un3

  See also: Palmer, E.L. Fieldbook of Natural History (No. A81)

ATLASES

  General

      See: Ingrao, H.C. & Kasparin, E. Photographic Star Atlases. Sky & Telescope 34:284-7, 1967.

E50. Barton, S.G. & Barton, W.H. Jr. Guide to the Constellations. McGraw-Hill, 1943. 523.89 B28g

E51. Boss, B. General Catalogue of 33,342 Stars for the Epoch 1950. Wash., Carnegie Institution of Washington. 1937. 5v. (Carnegie Institution Publication No. 468) (Reprinted 1962)
  See: Wa p. 67.

E52. Callataÿ, V. de. Atlas of the Sky. St. Martin's, 1957. q523.89 C13aEj
  See: Wa p. 65.

E53. Celestial Handbook. Celestial Handbook Pub., 1966- (8 sections projected)
  See: Sky & Telescope 31:362, 1966.

E54. Howard, N.E. The Telescope Handbook and Star Atlas. Crowell, 1967.
See: Sky & Telescope 35:174, 1968; Choice 4:1104, 1967.

E55. Levitt, I.M. & Marshall, R.K. Star Maps for Beginners. Simon & Schuster, 1964.

E56. Menzel, D.H. A Field Guide to the Stars and Planets. Houghton Mifflin, 1964. 523 M529f
See: Sky & Telescope 29:36-7, 1965; A.L.A. Booklist 60:944, 1964.

E57. Norton, A.P. & Inglis, J.G. Star Atlas and Reference Handbook (epoch 1950) for Students and Amateurs. 15th ed. Gall, 1964. (reprinted frequently) 523.89 N82s
See: Wi-EB24.

E58. Palomar Sky Atlas. Palomar Observatory, 1955-1960. (A limited edition of approximately 1800 photographic plates)
See: Astronomical Journal 61:336-7, 1956.

E59. Sandage, A. The Hubble Atlas of Galaxies. Wash., Carnegie Institution of Washington, 1961. (Carnegie Institution of Washington Publication No. 618) q 523.85 Sa5h

E60. Smithsonian Institution. Astrophysical Observatory. Star Catalog. G.P.O., 1966. 4v. 523.8908 Sm6s
See: Science 152:1610, 1966; Science News 89:297, 1966.

E61. Vehrenberg, H. Atlas of Deep Sky Splendors. Sky Pub., 1967.
See: Sky & Telescope 34:252-4, 1967.

E62. Vehrenberg, H. Photographic Star Atlas. Sky Pub. 1962-64.
See: Sky & Telescope 29:168-9, 1965.

E63. Sky and Telescope. Sky Pub. v.1, 1941-   . 520.5 SKA
See: Wi-EB9.

Special

E64. Alter, D. ed. Lunar Atlas. Dover, 1968. (Reprint of work originally published in 1964 by North American Aviation Inc. in limited edition.)
See: Sky & Telescope 36:112, 1968.

E65. Alter, D. Pictorial Guide to the Moon. Updated and expanded edition. Crowell, 1967.
　　See: Lib. J. 92:4425-6, 1967.

E66. Barabashov, N. P. et al. eds. An Atlas of the Moon's Far Side, the Lunik III Reconnaissance. Interscience, 1961. 523.34 AklaEr
　　See: New Tech. Bks. 47:2, 1962.

E67. California Institute of Technology. Jet Propulsion Laboratory. Ranger VII Photographs of the Moon. NASA, 1964-65. 3v. q523.39 c128r

E68. Callatay, V. de. Atlas of the Moon. St. Martin's, 1964. q523.39 C13aEl
　　See: Science 147:1027, 1965; Lib. J. 89:4922, 1964.

E69. Hoffleit, D. Catalogue of Bright Stars. Yale Univ. Observatory, 1964. 523.89 Sch3c
　　See: Sky & Telescope 29:174, 1965.

E70. International Union of Geodesy and Geophysics. International Auroral Atlas. Edinburgh Univ. Pr., 1963. q538.768 In8i
　　See: Sky & Telescope 29:312-3, 1965.

E71. Kopal, Z. et al. Photographic Atlas of the Moon. Academic 1965. q523.39 K83p
　　See: Science 152:954-5, 1966.

E72. Kuiper, G. P. et al. Photographic Lunar Atlas. Univ. of Chicago Pr., 1960. Supplements, 1960-68. f523.39 K95p
　　See: Am. Scientist 53:387A, 1965; L. C. Info. Bull. 27:299-300, 1968.

E73. Vaucouleurs, G. de & Vaucouleurs, A. de. Reference Catalogue of Bright Galaxies. Univ. of Texas Pr., 1964. q520 T313 No. 1
　　See: Sky & Telescope 29:172, 1965; Science 147:1568, 1965.

E74. Zwicky, F. et al. Catalogue of Galaxies and Clusters of Galaxies. Calif. Inst. of Tech., 1961-63. 2v. q523.85 Z97c
　　See: Sky & Telescope 29:107-8, 1965.

DIRECTORIES

E75. Directory of Meteorite Collections and Meteorite Research. UNESCO. 1968.

E76. International Astronomical Union. Les Observatoires
astronomiques et les Astronomes. Brussels, Observatoire Royale de
Belgique, 1959.
   See: Wa p. 64.

See also: American Ephemeris and Nautical Almanac. (No. E49)

CURRENT INFORMATION; SELECTION AIDS

Consult:

   Astronomical Journal, v. 1, 1849-   .
   Mathematical Reviews, v. 1, 1940-   .
   Natural History, v. 1, 1900-   .
   Planetarium, v. 1, 1968-   .
   Review of Popular Astronomy, v. 1, 1951-
   Sky and Telescope, v. 1, 1941-   .

The earth sciences is that group of sciences that studies the constitution, structure, and the history of the earth, and the processes that
alter it.   It is usually subdivided into various fields of specialization
such as:

Atmospheric sciences: the sciences concerned with the physical
    and chemical properties, the composition, behavior, and pro-
    cesses of the planetary atmospheres of the solar system.
    From a scientific point of view the major subdivisions are:
    (a) Meteorology: that portion of the atmospheric sciences
    concerned with the physics and chemistry of the earth's
    atmosphere as a continuum.   This includes such sections as
    physical meteorology, dynamic meteorology, analysis and
    forecasting, climatology, instrumentation, and system design
    (see Federal Council for Science and Technology.   National
    Atmospheric Sciences Program.   Wash., 1967.   ICAS Report
    No. 11); (b) Aeronomy: that subdivision concerned with the
    physics and chemistry of the earth upper atmosphere.
Cosmology: the study of the origin of the earth.
Economic geology: the study of valuable mineral deposits.   This
    includes such subdivisions as engineering geology, mining
    geology, petroleum geology, etc.
Geochemistry: study of the chemical composition and changes in
    the composition of the earth.
Geodesy: concerned with measuring the form and size of the earth,
    locating points upon its surface, and describing its gravity
    field at every point upon its surface.
Geomorphology: a science that deals with the land and submarine
    relief features on the earth's surface.
Geophysics: the physics of the earth's surface.
Hydrology: that branch which treats of the storage and movement
    of water on the earth, the physical and chemical reaction of
    water with its environment, and the relation of water to
    living organisms.   This includes water as it occurs in lakes,
    streams, and underground streams.
Mineralogy: the systematic study of the physical characteristics
    of the minerals of the earth's crust.   This includes chemical,
    physical, descriptive, determinative, and crystallographic
    mineralogy.

79

Oceanography: the science that deals with the ocean and its
    phenomena. Includes principles of biological, chemical,
    physical, and geological marine science.
Paleontology: the study of plants and animals that have in-
    habited the earth. It is subdivided into such natural areas as:
    vertebrate, invertebrate, paleobotany, and micropaleontology.
Petrology: the scientific study of the rocks of the earth's crust.
    Petrography deals with the scientific description and classi-
    fication of rocks. Stratigraphy covers the classification,
    nomenclature, correlation, and interpretation of stratified
    rocks.
Seismology: the study of movements within the earth's crust.
Volcanology: the scientific study of volcanoes and volcanic
    phenomena.

GENERAL

Guides to the literature

F1. Geoscience Information Society. Geologic Field Trip Guidebooks
of North America: a Bibliography and Union List. Houston, Texas,
Phil Wilson, 1968.

F2. Howell, J.V. & Levorsen, A.I. Directory of Geological Material
in North America. 2d ed., Am. Geol. Inst., 1957. (NAS-NRC Pub.
556) 016.55 H83d
    See: Wi-EE64.

F3. Kaplan, S.R. ed. A Guide to Information Sources in Mining,
Minerals, and Geosciences. Wiley, 1965. 016.622 K14g
    See: New Tech. Bks. 50:305-6, 1965; Lib. J. 91:1148, 1966.

F4. Mason, B. Literature of Geology. N.Y., 1953. A550 M381
    See: Wi-EE1. Wa p. 111.

F5. Pearl, R.M. Guide to the Geologic Literature. McGraw-Hill,
1951. A550 P31g
    See: Wi-EE2. Wa p. 111 (note).

F6. U.S. Geological Survey. Publications of the Geological Survey,
1879-1961. Supplements (annual) 1962- . (Supplemented monthly
by "New Publications of the Geological Survey") A557 Xp
    See: Wi-EE27.

F7. Ward, D.C. III. Geologic Reference Sources. Univ. of Colo. Pr., 1967. (University of Colorado Studies: Series in Earth Sciences No. 5) 016.55 W21g

See also:
  Long, H.K. Bibliography of Bibliographies on the Geology of the States of the United States. <u>Geoscience Abstracts</u> 7:115-125, 1965.
  Smith, F.D. Jr. Developing a Coordinated Information Program for Geological Scientists in the United States. Am. Geol. Inst., 1967. (PB 177-290)

<u>Bibliographies, Indexes, Abstracts</u>

F8. Annotated Bibliography of Economic Geology. Economic Geology Pub., v.1, 1929- A553 An78
  See: Wi-EE4.

F9. Bibliographie des sciences de la terre. Orléans, Bureau de recherches géologiques et minières. No. 1, 1968- . (Sections: A, Minéralogie et Géochimie; B, Gitologie et Economie Minière; C, Roches sedimentaires; E, Stratigraphie et Géologie; F, Techonique et Géophysique; G, Hydrogéologie et Géologie de l'Ingénieur; H, Paleontologie) 016.55 B471

F10. Bibliography and Index of Geology. Geological Society of America, v.33, 1969- . Monthly with an annual index.
  Successor to: Bibliography and Index of Geology Exclusive of North America, v.1-32, 1934-68. A550 G291b
  See: <u>Geotimes</u> 13:26, (Dec.) 1968. Wi-EE25.

F11. Bibliography of North American Geology. Wash., U.S.G.S. 1896- . (Issued as U.S.G.S. bulletins. Titles vary) A557.9 N53g
  See: Wi-EE24. Wa p. 115.

F12. Abstracts of North American Geology. Wash., G.P.O. v.1, 1966- . 557.05 AB

F13. Catalog of the United States Geological Survey Library, Department of the Interior. G.K. Hall, 1965. 25v. q016.55 Un3c

F14. Corbin, J.B., comp. An Index of State Geological Survey Publications Issued in Series. Scarecrow, 1965. 016.557 XC

F15. Earth Science Reviews. Elsevier. v.1, 1966-   550.5 EAR
  See: Coll. & Res. Libs. 28:66, 1967.

F16. Geoscience Abstracts. v.1, 1953-66. (Title varies) 550.5
GEAB
  See: Wi-EE34. Wa p. 110.

F17. Margerie, E. Catalogue des Bibliographies geologique ...
Paris, Gauthier-Villars, 1896. 016.55 M33c
  See: Wi-EE21. Wa p. 111.

F18. Mathews, E.B. Catalogue of Published Bibliographies in
Geology, 1896-1920. Wash., National Research Council, 1923.
(NRC Bull. 36) 500 N213 No. 36.
  See: Wi-EE22. Wa p. 111.

F19. Meisel, M. Bibliography of American Natural History ...
Brooklyn, Premier, 1924-29. 3v. (Reprinted by Hafner, 1967)
A570 M47b
  See: Lib. J. 93:65, 1968. Wi-EC23.

F20. Société Géologique de France. Bibliographie des Sciences
géologiques. 1923-60. A550 Solb
  See: Wi-EE6.

F21. U.S.G.S. Guide to Indexing Bibliographies and Abstract
Journals of the U.S. Geological Survey. Wash., 1967.

F22. U.S.G.S. Serial Publications Commonly Cited in Technical
Bibliographies of the U.S. Geological Survey. Wash., 1967.

F23. U.S. Library of Congress. United States IGY Bibliography,
1953-1960; an Annotated Bibliography of United States Contributions
to the IGY and IGC (1957-59). National Research Council, 1963.
(NRC Pub. No. 1087) 016.551 Un3u
  See: Wi-EE28.

F24. Zentralblatt für Geologie und Paläontologie. Stuttgart,
Schweizerbart'sche Verlags. 1950-   550.5 ZENG
  See: Wi-EE37. Wa p. 111.

See also:

  International Catalogue of Scientific Literature. H: Geology.
    (No. A16)
  Bulletin Signalétique. 10 & 11: Sciences de la terre. (No. A18)
  Referativnyi Zhurnal. Geologiya (No. A19)

Reviews & Surveys

F25. Advances in Geology. Academic, v.1, 1965- .

F26. Annals of the International Geophysical Year. Pergamon, v.1, 1957 -
  See: Wa p. 116.

F27. Chemical Society, London. Annual reports on the Progress of Chemistry. v.1, 1904- . 540.6 CH

F28. International Geology Review. Wash., American Geological Institute, v.1, 1959- . 550.5 IN
  See: Wa p. 110.

F29. International Conference on the Earth Sciences, Cambridge, Mass., 1964. Advances in Earth Science: Contributions. M.I.T. Pr., 1966. 550.8 In8a

Histories

F30. Adams, F.D. Birth and Development of the Geological Sciences. Williams & Wilkins, 1938. (Reprinted by Dover, 1954) 550.9 Adlb

F31. Geological Society of America. Geology 1888-1938. G.S.A., 1941. 550 G292g

F32. La Rocque, A. Contributions to the History of Geology. Ohio State University, Department of Geology. 1960-64. 3v. 550.9 L32c
  See: Wi-EE66a.

F33. Manning, T.G. Government in Science: the U.S. Geological Survey, 1867-1894. Univ. of Kentucky Pr., 1967. 557 Xman
  See: Science 159:415, 1968.

F34. Margerie, E. de. Critique et Géologie. Contribution à l'Histoire des Sciences de al Terre. Paris, Gauthier-Villars, 1943-48. 4v. 550.8 M33c

F35. Mather, K.E. & Mason, S.L. Source Book in Geology. McGraw-Hill, 1939. (Reprinted by Hafner, 1964) 550.8 M42s

F36. Mather, K.F. Source Book in Geology, 1900-1950. Harvard Univ. Pr., 1967. 550.8 M42so
  See: Science 158:898, 1967; Choice 4:1104, 1967.

F37. Merrill, G. P. The First One Hundred Years of American Geology. Yale Univ. Pr., 1924. (Reprinted by Hafner, 1964) 550 M55f

F38. White, G. W. Annotated Bibliography for History of Geology. Urbana, Ill. 1964.

F39. White, G. W., ed. Contributions to the History of Geology series. Hafner, 1968-   . (Reprints of classical works in geology)

See also:
   Ireland, H.A. History of the Development of Geologic Maps. Geological Society of America. Bulletin 54:1227-80, 1943.
   White, G. W. Early State Geological Survey Publications. Stechert-Hafner Book News 23:21-3, 1968.

Dictionaries & Encyclopedias

General

F40. Allbritton, C.C. Jr. ed. The Fabric of Geology. Freeman, 1964. 550.8 Allf
 See: Science 148:810, 1965.

F41. American Geological Institute. Glossary of Geology and Related Sciences; a Cooperative Project. 2d ed., Wash., 1960. (also available as 1st ed., 1957, plus Supplement, 1961) New edition in preparation (Geotimes 13:14, (May-June) 1968) 550.3 Am3g
 Abridged edition: Dictionary of Geological Terms. Doubleday, 1962. 550.3 Am3g 1962.
 See: Wi-EE42. Wa p. 112.

F42. Challinor, J. Dictionary of Geology. 3d ed. Oxford, 1967. 550.3 C35d
 See: Choice 4:1223, 1968. Wi-EE44. Wa p. 112.

F43. International Geophysical Year Special Committee. Annals of the International Geophysical Year. Pergamon, 1957-   . 551 In8a

F44. Larousse Encyclopedia of the Earth. Prometheus, 1961. q550.3 B462tEb
 See: Science 135:783, 1962.

F45. Lehrbuch der Angewandten Geologie. Stuttgart, Enke, 1961-   . 550 B451
 See: Science 135:913-4, 1962.

F46. Lexicon of Geologic Names of the United States for 1936-1960.
Wash., G.P.O., 1966. 3v. (U.S.G.S. Bulletin 1200) 557 Xw
  Earlier compilations published as the Surveys Bulletins 191, 769,
826, 896, 1056A and 1056B.

F47. Nelson, A. & Nelson, K.D. Dictionary of Applied Geology,
Mining, and Civil Engineering. Philosophical, 1967. 550.3 N33d

F48. Rice, C.M. Dictionary of Geological Terms, Exclusive of
Stratigraphic Formations and Paleontologic Genera and Species.
Edwards, 1940. Reprinted with addenda, 1961. 550.3 R36d
  See: Wi-EE49. Wa p. 112.

  See also:
    Moore, W.G. A Dictionary of Geography. 4th ed. Praeger,
    1967
      See: Lib. J. 92:3027, 1967.
    Stamp, D. ed. Dictionary of Geography. Wiley, 1966.
      See: Am. Scientist 55:106A, 1967.

    Foreign language

F49. Davies, G.M. French-English Vocabulary in Geology and
Physical Geography. Van Nostrand, 1932. 550.3 D28f
  See: Wi-EE52. Wa p. 112.

F50. Huebner, W. Geology and Allied Sciences. German-English.
Veritas, 1939. 550.3 H87g
  See: Wi-EE53. Wa p. 112 (note).

F51. Burgunker, M. Russian-English Dictionary of Earth Sciences.
Telberg, 1961. 551.03 B91r
  See: Wi-EE58.

F52. Sofiano, T.A. Russko-angliiskii Geologicheskii Slovar'.
Moscow, Fizmatgiz, 1960. American Supplement, Telberg, 1961.
550.3 An46
  See: Wi-EE60. Wa p. 113.

F53. Telberg, V.G. Russian-English Dictionary of Geological Terms,
Telberg, 1964.
  See: Wi-EE61.

F54. Royal Geological and Mining Society of the Netherlands.
Geological Nomenclature. Heinman, 1960. (Dutch, English,
French, German) 550.3 G29g
  See: Wi-EE55.

Handbooks & Tables

F55. Compton, R.R. Manual of Field Geology. Wiley, 1962.
550.7 C73
  See: New Tech. Bks. 47:81, 1962.

F56. Handbook of PhysicalConstants. Rev. ed. Geol Soc. Am.,
1966. (G.S.A. Memoir 97) 508 N195h
  See: Physics Today 20:77-8, 1967.

F57. National Academy of Sciences. Catalogue of Data in World
Data Center A. Wash., 1961-

Maps

   See:
      U.S.G.S. Publications of the Geological Survey. (No. F6)
      Bibliography and Index of Geology. (No. F10)
      Bibliography of North American Geology. (No. F11)

   Bibliographies, Indexes, etc.

F58. Porter, P.W. A Bibliography of Statistical Cartography.
Univ. of Minn., 1964. 016.5268 P83b

F59. Watkins, J.B. Selected Bibliography of Maps in Libraries;
Acquisition, Classification, Cataloging, Storage, Uses. Rev. ed.
Syracuse Univ. Libs, 1967. 016.912 W32s

F60. U.S. Geological Survey. A Descriptive Catalog of Selected
Aerial Photographs of Geologic Features of the United States.
G.P.O., 1968. (U.S.G.S. Professional Paper 590).
  See: New Publications of the Geological Survey, Sept. 1968.

F61. U.S. Geological Survey. Index Map of the Subterrestrial
Hemisphere of the Moon. G.P.O., 1962-
  See: New Publications of the Geological Survey, June 1968.

F62. U.S. Geological Survey. Index to Geologic Mapping in the
United States. 1947-   (A series of state index maps)

F63. U.S. Geological Survey. Index to Topographic Maps of the United States. 1935- . (A series of state index maps)

F64. U.S. Geological Survey. Status of Aerial Photography in the United States. 1961.

F65. U.S. Geological Survey. Status of Geologic Mapping in the United States, Territories, and Possessions. 5th ed., 1956.

F66. U.S. Geological Survey. Status of Topographic Mapping in the United States. 1961.

F67. U.S. Geological Survey. Transcontinental Geophysical Maps series. Wash., 1968- .
  See: New Publications of the Geological Survey, Nov. 1968. p. 11.

F68. U.S. National Aeronautics and Space Administration. Earth Photographs from Gemini III, IV, and V. Wash., G.P.O., 1967. (NASA Sp 129) q525 Un3e
  See: Sky and Telescope 35:40-1, 1968.

    General

F69. Birch, T.W. Maps, Topographical and Statistical. 2d ed. Clarendon, 1964. 526.98 B53m

F70. Blyth, F.G.H. Geological Maps and Their Interpretation. London, Arnold, 1965. q550.2 B62g

F71. Simpson, B. Geological Maps. Pergamon, 1968.

GEOCHEMISTRY

F72. Geochemical Prospecting Abstracts. Wash., U.S.G.S., 1953- . (U.S.G.S. Bulletins 1000A, 1000G, 1098B) Irregular. 557 Xb

F73. Handbook of Geochemistry. Springer-Verlag, 1969- .

F74. Researches in Geochemistry. Wiley, 1959-67. 2v. 551.13 Ab3r
  See: Science 159:867, 1968.

F75. U.S. Geological Survey. Analytical Methods Used in Geochemical Exploration by the U.S. Geological Survey. 1963. (U.S.G.S. Bull. 1152) 557 Xb

GEOMORPHOLOGY

F76. Geo Abstracts. London, London School of Economics, v.1, 1966-
Section A: Geomorphological Abstracts. (Continues Geomorphological Abstracts, v.1-6, 1960-5) 551.405 GE
  See: Wa p. 119.

F77. Chorley, R.J. et al. The History of the Study of Landforms.
Wiley, 1964- . 551.4 C55h
  See: Wa p. 123.

F77a. Fairbridge, R.W., ed. The Encyclopedia of Geomorphology.
Reinhold, 1968. (Encyclopedia of Earth Sciences Series v.3)
  See: New Tech. Bks. 54:129, 1969.

F78. U.S. Military Academy, West Point. Atlas of Landforms.
Wiley, 1966. 551.4 Un33a
  See: Choice 4:28, 1967.

GEOPHYSICS

F79. Advances in Geophysics. Academic, v.1, 1952- . 551 Ad95
  See: Science 117:431, 1953. Wa p. 116.

F80. Geophysical Abstracts. Wash., U.S.G.S., No. 1, 1929-
551.05 UNIB
  See: Wi-EE36. Wa p. 115.

F81. List of Journals Commonly Cited in Geophysical Abstracts.
U.S.G.S., 1961- . 016.55105 L69

F82. Geophysical Directory. Houston, Texas. v.1, 1946- .
  See: Wi-EE63.

F83. International Dictionary of Geophysics. Seismology, Geomagnetism, Aeronomy, Oceanography, Geodesy, Gravity, Marine geophysics, Meteorology, the Earth as a Planet and its Evolution. Pergamon, 1967. 2v. and atlas. 551.03 In8
  See: Science 160: 755, 1968. Choice 5:1284, 1968.

F84. International Union of Geodesy and Geophysics. Abstracts of Papers. v.1, 1963- . 526.106 In8a

F85. Physics and Chemistry of the Earth. Pergamon, v.1, 1957- . 551 P56
  See: Science 125:891, 1957. Wa p. 116.

F86. U.S. Air Force. Cambridge Research Laboratories. Handbook of Geophysics and Space Environments. McGraw-Hill, 1965. 551 Un32h
  See: Am. Scientist 54:338A-9A. 1966. Wa p. 116.

  See also: Referativnyi Zhurnal. Geofizika. (No. A19)

HYDROLOGY

F87. Annotated Bibliography on Hydrology and Sedimentation.
G.P.O., 1952- . (Title varies; succeeds Bibliography of Hydrology)
016.55146 Am3ann
  See: Wi-EE76.

F88. Chow, V.T. ed. Handbook of Applied Hydrology; a Compendium of Water Resources Technology. McGraw-Hill, 1964. 551.89 C458h
  See: Science 148:219, 1965. Wi-EE77.

F89. Frey, D.G. ed. Limnology in North America. Univ. of Wis., 1963. 551.48 F8971
  See: Science 141:420-1, 1963.

F90. Hydrata. Urbana, Ill., American Water Resources. v.1, 1965- . 016.55146 H999
  See: Science 148:1449, 1965.

F91. Traité de Glaciologie. Paris, Masson, 1964-65. 2v.
  See: Science 156:1475, 1967.

F92. U.S. Library of Congress. Bibliography on Snow, Ice, and Permafrost with Abstracts. v.1, 1951- . (title varies)
551.573 Un34b

F93. U.S. Office of Water Resources Research. Water Resources Research Catalog. Wash., G.P.O., 1965- . 016.55149 Un35w
  See: Geotimes 9:24, (May/June) 1965.

F94. U.S. Office of Water Resources Research. Water Resources Thesaurus. Wash., G.P.O., 1966. 025.36 U574w
  See: Am. Doc. 18:209-15, 1967.

METEOROLOGY

Bibliographies, Indexes, Abstracts

F95. Meteorological and Geoastrophysical Abstracts. Am. Metero-
logical Soc., v.1, 1950   . (title varies) 551.505 META
  Note bibliography in monthly issues, particularly those on out-
standing books in the January and February 1960 issues.
  See: Wi-EE82. Wa p. 127.

F96. Meteorological and Geoastrophysical Titles. Am. Meteorological
Soc. v.1- , 1961-   . (experimental journal) 551.505 MET
  See: Wa p. 127.

F97. Kiss, E. ed. Bibliography on Meteorological Satellites (1952-
1962). U.S. Weather Bureau, 1963. 016.551591 M56b
  See: Wi-EE80.

F98. U.S. Weather Bureau. Selective Guide to Published Climatic
Data Sources Prepared by U.S. Weather Bureau. Wash., 1963.
016.5515 Un32s
  See: Wi-EE100.

Histories

F99. Popkin, R. The Environmental Science Service Administration.
Praeger, 1967. 353 P8le
  See: Science 160:755-6, 1968.

F100. Witnah, D.R. History of the United States Weather Bureau.
Univ. of Ill. Pr., 1961.

Dictionaries, Encyclopedias, Descriptive works

F101. American Meteorological Society. Compendium of Meteorology.
Boston, 1951. 551.5 Am35c
  See: Nature 170:814-5, 1952. Wi-EE89.

F102. Annals of the IQSY. MIT Pr., v.1, 1968-   . 523.2 Sp3a
  See: Sky & Telescope 36:107-8, 1968. Choice 5:1157, 1968.

F103. Glossary of Meteorology. Am. Meteorological Soc., 1959.
551.503 G516
  See: Science 131:222, 1960. Wi-EE83 Wa p. 128.

F104. Great Britain. Meteorological Office. Meteorological Glossary. 4th ed. London, H.M.S.O., 1963. 551.503 G79m
  See: Wi-EE84. Wa p. 128.

F105. Brazol, D. Dictionary of Meteorological and Related Terms. English-Spanish, Spanish-English. Buenos Aires, Hachette, 1955. 551.503 B73d
  See: Wi-EE88.

F106. World Meteorological Organization. International Meteorological Vocabulary. Geneva, 1966.
  See: Wi-1EE13a.

  See also:
    Fairbridge, R.W., ed. The Encyclopedia of Atmospheric Sciences and Astrogeology. (No. E25)

Handbooks & Tables

F107. Conway, H.M. Jr., ed. The Weather Handbook; a Summary of Weather Statistics for Principle Cities throughout the United States and Around the World. Conway Pub., 1963. 551.5 C773w
  See: Wi-EE96.

F108. Smithsonian Institution. Smithsonian Meteorological Tables. 6th ed. Wash., 1951. Reprinted, 1963. 506 Sm6m
  See: Nature 170:175, 1952. Wa p. 128.

F109. U.S. Weather Bureau. Climates of the States. Wash., G.P.O., 1959-   . (Climatology of the United States No. 60) 551.5 Un3clim

F110. U.S. Weather Bureau. Climatological Data for the United States by Sections. Wash., v.1, 1914-   (Monthly with annual summaries) 551.05 UNIC
  See: Wi-EE97.

F111. U.S. Weather Bureau. Climatological Data; National Summary. Wash., v.1, 1950. (Monthly with annual summary)

F112. U.S. Weather Bureau. World Weather Records. Wash., v.4, 1959-   . (earlier volumes issued by Smithsonian Institution) 551.5 W895
  See: Wi-EE95.

F113. U.S. Environmental Data Service. Daily Weather Maps,
Weekly Series. Wash., G.P.O., 1968-   (Continues: U.S. Weather
Bureau. Daily Weather Map. 1945-   .)

F114. Weatherwise; a Magazine about Weather. Am. Meteorological
Society, v.1, 1948-   . 551.505 WEA

See also appropriate sections in general almanacs.

MINERALOGY

Bibliographies, Indexes, Abstracts

F115. Mineralogical Abstracts. London, Mineralogical Society,
v.1, 1920-   . Issued as a supplement to Mineralogical Magazine
until 1959; a separate publication beginning with v.14, 1959-   .
549.05 MIN
   See: Wi-EE102. Wa p. 108.

F116. Zeitschrift für Kristallographie ... Leipzig, Engelmann.
v.1, 1877-   . 549.05 ZE

F117. Zentralblatt für Mineralogie. Stuttgart, Schweizerbart'sche.
1830-   . (title varies) 549.05 ZEN
   See: Wi-EE105. Wa p. 108.

See also: Kaplan, S.R., ed. A Guide to Information Sources in
Mining, Minerals, and Geoscience. (No. F3)

Dictionaries & Encyclopedias

F118. Chamber's Mineralogical Dictionary. Chemical Pub., 1948.
549.03 C35
   See: Wi-EE108. Wa p. 109.

F119. Pearl, R.M. Gems, Minerals, Crystals and Ores; the Col-
lector's Encyclopedia. Odyssey, 1964. Reprinted by Golden Pr.,
1967. 549 P31g
   See: Booklist 60:367, 1964.

F120. Bradley, J.E.S. & Barnes, A.C., comps. Chinese-English
Glossary of Mineral Names. Plenum, 1963. 549.03 B72c
   See: Wi-EE107. Wa p. 109 (note).

F121. Börner, R. Minerals, Rocks, and Gemstones. Edinburgh, Oliver & Boyd, 1966. A reprint of the 1962 translation of the German edition (1938) with additional plates and references. 549 B632wEm

F122. Bragg, Sir L. & Claringbull, G. F. Crystal Structures of Minerals. London, Bell, 1965. (v. 4 of the series "The Crystalline State")

F123. Dana, J.D. & Hurlbut, C.S. Manual of Mineralogy. 17th ed. Wiley, 1959. 549 D19m
  See: Sub. Bks. Bull. 56:243, 1959. Wa p. 108.

F124. Dana, J.D. & Dana, E.S The System of Mineralogy. 7th ed. Wiley, 1944- . (v.1, 1944; v.2, 1951; v.3, 1962; v.4-5 in preparation) 549 D193.
  See: Nature 199:1126, 1963; Science 139:821, 1963. Wi-EE113. Wa p. 108.

F125. Deer, W.A. Rock-forming Minerals. Longmans, 1962-3. 5v. 549 D36r
  Condensation published as: An Introduction to the Rock-forming Minerals. Longmans, 1966. (See: Am. Scientist 55:113A-4A, 1967)
  See: Wi-EE114. Wa p. 108.

F126. Fritzen, D.K. The Rock-Hunter's Field Manual; a Guide to Identification of Rocks and Minerals. Harper, 1959. 549.1 F91r
  See: Sub. Bks. Bull. 57:411+, 1959.

F127. Gleason, S. Ultraviolet Guide to Minerals. Van Nostrand, 1960. 549.1 G47u
  See: Sub. Bks. Bull. 57:416, 1961.

F128. Liddicoat, R.T. Handbook of Gem Identification. 7th ed. Los Angeles, Gemological Institute of America, 1966. 553.8 L61h

F129. MacFall, R.P. Gem Hunter's Guide; How to Find and Identify Gem Minerals. 4th ed. Crowell, 1969. 553.8 M16g
  See: Lib. J. 94:2326, 1969.

F130. Pough, F.H. Field Guide to Rocks and Minerals. 3d ed. Houghton, 1960. 549.1 P86f
  See: Tech. Bk. Rev. Index 27:116, 1961.

F131. Ransom, J.E. A Range Guide to Mines and Minerals: How and Where to Find Valuable Ores and Minerals in the U.S. Harper & Row, 1964. 549.973 R17r
  See: Wi-EE118.

F132. Sinkankas, J. Van Nostrand's Standard Catalog of Gems.
Van Nostrand, 1968. 553.8 Si6v
  See: Booklist 65:142, 1968.

F133. U.S. Bureau of Mines. Mineral Facts and Problems. Wash.,
G.P.O., 1965. (also available in separate chapters) 338.2 Un35min
  See: Wa p. 379.

F134. U.S. Bureau of Mines. Minerals Yearbook. Wash., G.P.O.,
1933-   . Annual. 1963-65 issued in 4v.; 1966 in3v. 553 Un32m
  See: Wi-EE120. Wa p. 378.

F135. Uytenbogaardt, W. Tables for Microscopic Identification of
Ore Minerals. Princeton Univ. Pr., 1951. Reprinted by Hafner,
1968. 549.1 Uy8t
  See: Choice 6:198, 1969.

F136. Vanders, I. & Kerrs, P.F. Mineral Recognition. Wiley,
1967. 549 V28m
  See: Lib. J. 91:1634, 1967.

F137. Zussman, J., ed. Physical Methods in Determinative Mineralogy. Academic, 1967. 549.12 Z89p

OCEANOGRAPHY

  Bibliographies, Indexes, Abstracts, Reviews

F138. Bibliographia Oceanographica. Venice, v.1-29, 1928-56.
A591.92 B471

F139. Current Bibliography for Aquatic Sciences and Fisheries.
Winchester, England, Taylor & Francis, 1964-   016.57492 C936
  See: Science 148:786-7, 1965. Wa p. 169, 318.

F140. Deep-Sea Research and Oceanographic Abstracts. Pergamon,
v.1, 1953/54-   . (Title varies) 551.40605 DG

F141. Fogel, L.J. Composite Index to Marine Science and Technology.
San Diego, Calif., Alfo Pub., 1966. 016.55146 F68c
  See: Lib. J. 93:2640, 1968.

F142. National Research Council. Oceanography Information Sources.
Wash., 1966. (NAS-NRC Pub. 1917) 016.55146 N21o

F143. Oceanic Abstracts. La Jolla, Calif., Oceanic Library and
Information Center. v.1, 1966- . (annual) 551.4605 OCA
  v. 1-2: State of the Art-Instrumentation. (annotated bibliography)

F144. Oceanic Index. La Jolla, Calif., Oceanic Research Institute.
v.1, 1964- .
  Title and format of this coordinate index vary. Issued loose-leaf
in 1964-67, and replaced later by bound volume. In 1968 the volume
was issued in 2 parts: Citation Journal with Abstracts (monthly);
Keytalpha Index (each quarterly issue cumulated, terminating in an
annual)

F145. Oceanography, the Weekly of the Ocean. Wash., American
Aviation, v.1, 1966- . 551.4605 OCEA

F146. Progress in Oceanography. Pergamon, v.1, 1963-   551.46
P63
  See: Science 144:987, 1964. Wa p. 125 (note).

Dictionaries, Encyclopedias, Descriptive works

F147. Fairbridge, R.W. The Encylopedia of Oceanography. Reinhold,
1966. 551.4603 F15e
  See: Am. Scientist 55:126A-7A, 1967; Lib. J. 92:996, 1967.

F148. Hunt, L.M. & Groves, D.G., eds. A Glossary of Ocean
Science and Undersea Technology Terms. Compass Pub., 1965.
  See: Wi-1EE18.

F149. Jerlov, N.G. Optical Oceanography. Elsevier, 1968.
55146 J47o

F150. Riley, J.P. & Skirrow, G. eds. Chemical Oceanography.
Academic, 1965- . 551.46 R45c
  See: New Tech. Bks. 51:56 and 88, 1966.

F151. The Sea; Ideas and Observations on Progress in the Study of
the Sea. Interscience, 1962-3. 3v. 551.46 Se11
  See: Science 140:644-5, 1963; 142:215-6, 1963.

F152. U.S. Naval Oceanographic Office. Glossary of Oceanographic
Terms. 2d ed. Wash., 1966. (Special Pub SP-35) 551.4603 Un3g.
  See: Wi-1EE19.

Handbooks & Tables

F153. Handbook of Oceanographic Tables, 1966. Wash., G.P.O., 1967. 551.46 B47h

F154. IGY World Data Center A: Oceanography. Catalogue of Data in World Data Center A. Wash., 1957/1963- . Supplements, 1964- 551.46 C28

Directories

F155. International Directory of Oceanographers. 4th ed. NAS-NRC, 1964. 551.46 In86
 See: Wi-EE124.

F156. World Directory of Marine Laboratories. Reinhold, 1963.

PALEONTOLOGY

Bibliographies, Indexes

F157. Andrews, H.N. Jr. Index of Generic Names of Fossil Plants, 1820-1950. Wash., U.S.G.S., 1955. (U.S.G.S. Bull. 1013) 557 Xb No. 1013
 Extensive bibliography p. 13-98.
 See: Wi-EE131. Wa p. 134.

F158. Ellis, B.F. & Messina, A.R. Catalogue of Foraminifera... N.Y., American Museum of Natural History, 1940- . (looseleaf) Continued by looseleaf supplements. q593.12 E15c
 See: Wi-EE132, 1EE20.

F159. Romer, A.S. et al. Bibliography of Fossil Vertebrates Exclusive of North America, 1509-1927. Geol. Soc. Am., 1962. 2v. (G.S.A. Memoir 87) 550 G292m No. 87
 See: Wi-EE128. Wa p. 135.

F160. Bibliography of Fossil Vertebrates. N.Y., Geol. Soc. Amer. 1902- . Literature to 1900 comp. by O.P. Hay (U.S.G.S. Bull. 179), 1902. 557 Xb
　　1902-1927 comp. by O.P. Hay (Carnegie Inst. Pub. no. 390, v.1-2),
　　　　1929-1930.
　　1928-1933 comp. by C.L. Camp et al. (GSA Spec. Paper 27), 1940.
　　1934-1938 comp. by C.L. Camp et al. (GSA Spec. Paper 42), 1942.

1939-1943 comp. by C. L. Camp et al. (GSA Memoir 37), 1949.
1944-1948 comp. by C. L. Camp et al. (GSA Memoir 57), 1953.
1949-1953 comp. by C. L. Camp et al. (GSA Memoir 84), 1961.
1954-1958 comp. by C. L. Camp et al. (GSA Memoir 92), 1964
See: Sci. Info. Notes 7:18, (June/July) 1965. Wi-EE127.

Comprehensive works

F161. Fossilium Catalogus. Berlin, Junk, v.1, 1913-   .
See: Wi-EE134, 1EE21. Wa p. 133.

F162. Principles of Zoological Micropalaeontology. Pergamon, 1963-
   This is a translation of the German edition of: Pokorný, V. Zaklady
Zoologiscke Mikropaleontologie, a Czechoslovakian publication)
560 P752Ea
See: Am. Scientist 52:230A-1A, 1964.

F163. Seward, A.C. Fossil Plants, for Students of Botany and
Geology. Cambridge Univ. Pr., 1898-1919. 4v. Reprinted Hafner,
1963. 561 Se8f

F164. Shimer, H.W. & Shrock, R.R. Index Fossils of North America.
Wiley, 1944. 562 Sh6i
See: Wi-EE136.

F165. Traité de Paléobotanique. Paris, Masson, 1964-   . (9v. projected) 561 B66t
See: Science 148:619-29, 1965; 160:1442, 1968.

F166. Traite de Paleontologie, publie sous la direction de Jean
Piveteau. Paris, Masson, 1952-   . 560 P68t
See: Science 117:429, 1953; 119:698, 1954; 129:1020, 1959.
Wi-EE141. Wa p. 133.

F167. Treatise on Invertebrate Paleontology. N.Y., Geol. Soc. Am.
1953- . 562 J66c
See: Am. Scientist 55:83A, 1967. Wi-EE135, 1EE22. Wa p. 134.

Handbooks

F168. Fenton, C.L. & Fenton, M.A. The Fossil Book. Doubleday,
1958. 560 F36f
See: Wi-EE138. Wa p. 133.

F169. Kummel, B. & Raup, D.M. eds. Handbook of Paleontological Techniques. Freeman, 1965. 560.18 K96h
  See: Science 148:354, 1965. Wi-1EE23.

F170. Oakley, K.P. Frameworks for Dating Fossil Man. Aldine, 1965. 571 Oa4f

F171. Ransom, J.E. Fossils in America. Harper, 1964. 560.973 R174f
  See: Wi-EE139.

F172. Rhodes, F.H.T. et al. Fossils, a Guide to Prehistoric Life. Golden, 1962. 560 R34p
  See: Wi-EE140.

  Directories

F173. Directory of Palaeontologists of the World. Hamilton, Canada, International Palaeontological Union. 1968. 560 D61

SEISMOLOGY, VOLCANOLOGY

F174. Bibliography of Seismology. Ottawa, Dominion Observatory, 1929-66.

F175. International Association of Volcanology. Catalogue of the Active Volcanoes of the World. Naples, 1951-   551.21 In84c
  See: Wi-EE149, 1EE24. Wa p. 117.

F176. U.S. Coast and Geodetic Survey. Earthquake History of the United States. Wash., G.P.O., 1958-61. 2v. 551.22 Un31ear
  See: Wi-EE146.

F177. U.S Coast and Geodetic Survey. United States Earthquakes. Wash., G.P.O., 1930-   . 551.22 L51p
  See: Wi-EE147.

CURRENT INFORMATION; SELECTION AIDS

  Current issues of:
    American Association of Petroleum Geologists. Bulletin, v.1,
      1917.
    Earth Science, v.1, 1946-   .
    Economic Geology, v.1, 1905-   .

GeoTimes, v.1, 1956-   .
Journal of Geological Education, v.1, 1951-   .
Journal of Geology, v.1, 1893-   .
GeoScience Information Society. Newsletter, 1967-

Publications of:
    Geological Society of America
    U.S. Bureau of Mines
    U.S. Geological Survey
    U.S. Hydrographic Office
    Smithsonian Institution.

SECTION G.   BIOLOGICAL SCIENCES

The biological sciences, often referred to as the life sciences, is
that group of sciences concerned with the study of living matter, its
structure, development, growth, distribution, functions, etc.
Division of the field may follow two fundamental principles:

1. Division according to the type of organism studied into the major
   disciplines:
   Botany: the study of plants
   Zoology: the study of animals.
   These divisions in turn may be subdivided according to species,
   e.g., entomology (insects), bryology (mosses), ornithology
   (birds).
2. Division according to aspects that are common to all types of
   organisms into such disciplines as:
   Morphology: the study of form and structure of organisms.
   Taxonomy: the classification of plants and animals.
   Anatomy: the study of the gross structure of an organism.
   Cytology: study of cells that compose the organism.
   Histology: microscopic study of tissues and organs.
   Physiology: study of the functions and activities of organisms.
   Biochemistry: study of the chemistry of life processes.
   Biophysics: concern with physical phenomena in biological systems.
   Pathology: study of the cause and effect of diseased organisms.
   Genetics: concerned with heredity and variation in living things.
   Embryology: the study of the early stages in the development of
     organisms.
   Ecology: study of interrelationships of organisms with their
     environment.

GENERAL BIOLOGY

   Guides to the literature

   See also:
   Blake, J.B. & Roos, C., ed. Medical Reference Books,
     1697-1966. (No. H1)

G1. Bottle & Wyatt, H. V., eds. The Use of Biological Literature. London, Butterworth, 1966; Archon, 1967. 016.574 B65u
  See: <u>Biblio. Soc. of Am.</u> 61:290, (July/Sept.) 1967; <u>Wilson Lib. Bull.</u> 42:220, 1967.

G2. Bourliére, F. Elements d'un Guide bibliographique du Naturaliste. Macon, Protat, 1940. Supplements, 1-2, Paris, Lechevalier, 1941.
  See: Wa p. 51.

G3. Kerker, A. E. & Murphy, H. T. Biological and Biomedical Resource Literature. Purdue Univ., 1968. 016.57 P9721

G4. Special Libraries Association. Information Sources for the Biological Sciences and Allied Fields. N.Y., 1961. 016.574 Sp3i
  See: Wa p. 142 (note).

Bibliographies, Indexes, Abstracts

G5. Altsheler, B. Natural History Index-Guide. 2d ed. Wilson, 1940. A570 A17n2
  See: Wa p. 51.

G6. Berichte über die gesamte Biologie. Abt. A: Berichte über die wissenschaftliche Biologie. v.1, 1926- . Abt. B: Berichte über die gesamte Physiologie unde experimental Pharmakologie. v.1, 1920- Abt. A 570.5 BE; Abt. B 612.05 BER

G7. Biological Abstracts. Philadelphia, Biological Abstracts, v.1, 1926- .
  Indexed in: BASIC. v.1, 1962- .
  Issued in the following sections: A: General Biology, 1939-62; B: Basic Medical Sciences, 1939-62; C: Microbiology, Immunology, Parasitology, 1939-62; D: Plant Sciences, 1939-62; E: Animal Sciences, 1939-62; F: Animal Production and Veterinary Science, 1942-53; G: Food and Nutrition Research, 1943-53; H: Human Biology, 1946-53; I: Cereal Products, 1947-53. 570.5 BIOA
  See: Wi-EC7. Wa p. 142.

G8. BioResearch Index. Philadelphia, Biological Abstracts. v.1, 1965- . (formerly BioResearch Titles) 574.05 BIORG
  See: Wi-1EC1.

G9. Biological & Agricultural Index. H. W. Wilson. v.50, 1964- . Continues Agricultural Index. v.1-49, 1916-64. A630 Ag829
  See: Wi-EC6. Wa p. 294 (note).

G10. British Museum (Natural History) Library. Catalogue of the Books, Manuscripts, Maps and Drawings in the British Museum (Natural History). London, 1903-40. 8v. Stechert-Hafner reprint, 1964. A500 B76c
   See: Wi-EA6. Wa p. 51.

G11. Current Contents: Life Sciences. Philadelphia, Institute for Scientific Information. v. 1, 1958-   (Title varies) 505 CUR

G12. International Abstracts of Biological Sciences. Pergamon, v. 1, 1954- . (title varies) Succeeded British Abstracts, Section AIII. 610.5 BRI
   See: Wi-ED16. Wa p. 142.

G13. BioScience. Wash., Am. Inst. Biol. Sci., v. 1, 1951-
   (Title varies) 570.6 AI

   See also:
      Bulletin Signalétique: Sections 12-17. (No. A18)
      International Catalogue of Scientific Literature: L, General
         Biology. (No. A16)
      Meisel, M. Bibliography of American Natural History. (No. F19)
      Referativnyi Zhurnal. Biologiya. (No. A19)

   Sections MEDICAL SCIENCES and AGRICULTURAL SCIENCES

Reviews & Surveys

G14. L'Annee Biologique. Paris, Fédération Francaise des Sociétés de Sciences Naturelles. v. 1, 1895-   570.5 ANN

G15. Ergebnisse der Biologie. Berlin, v. 1, 1926-   570 E238

G16. Federation of American Societies for Experimental Biology. Federation Proceedings. Wash., v. 1, 1958- . 610.6 FEDE

G17. Quarterly Review of Biology. Stony Brook, N.Y., Stony Brook Foundation, v. 1, 1926- . 570.5 Qu

G18. Society for Experimental Biology and Medicine. Proceedings. Utica, N.Y., v. 1, 1903/04- . 610.6 NES

G19. Survey of Biological Progress. Academic, v. 1, 1949-
570 Su79

102

G20. Viewpoints in Biology. London, Butterworth, v. 1, 1962-
574 V67

Histories

G21. Bradbury, S. The Evolution of the Microscope. Pergamon,
1967. 578.09 B72e
  See: BioScience 18:57, (Jan.) 1968.

G22. Dawes, B. A Hundred Years of Biology. Macmillan, 1952.
570.9 D32
  See: Wi-EC22. Wa p. 145.

G23. Gardner, E. J. History of Biology. 2d ed. Burgess, 1965.
574.09 G17h

G24. Nordenskiöld, E. The History of Biology. Knopf, 1928.
Reprinted Tudor, 1935 and 1960. (Translated by L. B. Eyre from the
Swedish original, Biologins historia, 1920-24. 3v.) 570.9 N75bE
  See: Wa p. 145.

G25. Singer, C. A History of Biology. 3d ed., Schuman, 1959.
(1st ed., 1931 has title: A Short History of Biology; American edition:
The Story of Living Things) 570.9 Si6h
  See: Wa p. 145 (note)

G26. Journal of the History of Biology. Harvard Univ. Pr., v. 1,
1968- . (semiannual) 574.05 HOH.
  See: Am. Scientist 56:478A-9A, 1968.

  See list of titles in:

    Bottle, R. T. & Wyatt, H. V. eds., The Use of Biological Litera-
      ture. p. 242-3. (No. G1)
    Medical Library Association. Medical Reference Works, 1679-
      1966. p. 104-5. (No. G4)

Dictionaries & Vocabulary lists

  General

G27. Abercrombie, M. et al. A Dictionary of Biology. 5th ed.
Penguin, 1966. 570.3 Ad37d
  See: Wi-1EC2. Wa p. 143.

G28. Compton's Dictionary of the Natural Sciences. Compton, 1966. 2v.
  See: Sub. Bks. Bull. 62:1053-6, 1966.

G29. Gray, P. The Dictionary of the Biological Sciences. Reinhold, 1967. 574.03 G79d
  See: Am. Scientist 56:90A-91A, 1968. Lib. J. 92:4491, 1967.

G30. Henderson, I.F. & Henderson, W.D. A Dictionary of Biological Terms. 8th ed. Van Nostrand, 1963. 503 H38d
  See: Wi-EC12. Wa p. 143.

G31. Jaeger, E.C. The Biologist's Handbook of Pronunciations. Thomas, 1960. 570.3 J17b
  See: Wi-EC14. Wa p. 144 (note).

G32. U.S. National Agricultural Library. Agricultural/Biological Vocabulary. Wash., 1967. 2v. Supplement, 1968- . 025.36 U5776ag
  See: Med. Lib. Assn. Bull. 56:535, 1968.

  Foreign language

G33. Artschwager, E.F. Dictionary of Biological Equivalents, German-English. Williams & Wilkins, 1930. 570.3 Ar7d
  See: Wa p. 144.

G34. Carpovich, E.A. Russian-English Biological and Medical Dictionary. N.Y., Technical Dictionaries, 1958. 570.3 K14r
  See: Chem. & Eng. News 37:86+, (Sept. 21) 1959. Wa p. 144.

G35. Dumbleton, C.W. comp. Russian-English Biological Dictionary. Plenum, 1965. 570.3 D89r
  See: Wa p. 144.

  Terminologies

G36. Borrer, D.J. A Dictionary of Word Roots and Combining Forms; Compiled from the Greek, Latin and other Languages, with Special Reference to Biological Terms and Scientific Names. Palo Alto, Calif., N-P Pub., 1960. 574.03 B64d

G37. Jaeger, E.C. Source-Book of Biological Names and Terms. 3d ed. Thomas, 1955. 570.3 J17s
  See: Wi-EC15. Wa p. 144.

G38. Woods, R.S   The Naturalist's Lexicon, a List of Classical Greek and Latin Words Used, or Suitable for Use, in Biological Nomenclature, with Abridged English-Classical Supplement. Pasadena, Calif., Abbey Garden Pr., 1944. Addenda, 1947. 570.3 W86n
    See: Wi-EC16. Wa p. 144.

    See also: Hough, J.N. Scientific Terminology (No. A59)

Encyclopedias

G39.   The Audubon Nature Encyclopedia.   Curtis Books, 1965.   12v.

G40.   Encyclopedia of the Life Sciences.   Doubleday, 1964-7.   8v.
    Series: v.1, The Living Organism; v.2, The Animal World; v.3, The World of Plants; v.4, The World of Microbes; v.5, The Human Machine: Mechanisms; v.6, The Human Machine: Disorders; v.7, The Human Machine: Adjustments; v.8, Man of Tomorrow.
    See: Science 150:1151-2, 1965; Am. Scientist 55:122A, 1967.

G41.   Gray, P. ed.   Encyclopedia of the Biological Science.   2d ed., Van Nostrand, 1969.   574.03 G79e
    For comments on 1st ed. see: Sub. Bks. Bull.   58:585-8, 1962; Science 134:93-4, 1961. Wi-EC18. Wa p. 143.

G42.   Shilling, C.W.   ed.   Atomic Energy Encyclopedia in the Life Sciences.   Saunders, 1964.   574.1903 Sh7a
    See: Chem. & Eng. News 42:60, (Nov. 16) 1964.   Wa p. 197 (note)

Comprehensive works

G43.   Abderhalden, E.   Handbuch der Biologischen Arbeitsmethoden. Berlin, Urban, 1920-1939.   13v.   612.01 Ab3h2

G44.   Biomedical Sciences Instrumentation.   Plenum, v.1, 1963- 574.19 N21b

G45.   Bittar, E.E. & Bittar, N. eds.   The Biological Basis of Medicine.   Academic, v.1, 1968-   (6v. projected)

G46.   Brachet, J. & Mirsky, A.E.   The Cell; Biochemistry, Physiology, Morphology. Academic, 1959-64.   6v.   576.3 B72c
    See: New Tech. Bks.   47:106-7, 1962.

G47. Giese, A.C. ed. Photophysiology. Academic, 1964. 2v. 574.191 G277p
  See: Science 144:1561-2, 1964.

G48. Physical Techniques in Biological Research. 2d ed., Academic, 1966- . 574 Os7p
  First edition, 1955-64, in 6v. was edited by G. Oster & A. W. Pollister.

Handbooks

  See also: Palmer, E.L. Fieldbook of Natural History. (No. A84)

G49. Altman, P.L. & Dittmer, D., eds. Biological Data Book. Wash., Federation of American Societies for Experimental Biology, 1964. (Revised edition of: Spector, W.S., ed. Handbook of Biological Data, 1956) 574 Sp3h
  See: Nature 206:971, 1965. Wi-EC19.

G50. American Institute of Biological Sciences. Biology Teachers' Handbook. Wiley, 1963. 574.07 Am3bi
  See: Science 143:668-70, 1964.

G51. Conn, H.J. Biological Stains: a Handbook on the Nature and Uses of the Dyes Employed in the Biological Laboratory. 7th ed. Geneva, Biotech Pubs., 1961. 578.6 C76b
  See: Wa p. 150.

G52. Gray, P. Handbook of Basic Microtechnique. 3d ed. McGraw-Hill, 1964. 578 G79h
  See: Choice 1:254, 1964.

G53. Gurr, E. Encyclopedia of Microscopic Stain. Williams & Wilkins, 1960. 578.64 G96e
  See: Science 132:614, 1960; Wa p. 150.

G54. Lenhoff, E.S. Tools of Biology. Macmillan, 1966.

G55. Scanga, F., ed. Atlas of Electron Microscopy; Biological Applications. Elsevier, 1964. q578.15 Sc63aE

G56. Stehli, G. The Microscope and How to Use It. Sterling, 1960. Reprinted by Cornerstone, 1963. 578 St32mE
  See: Sub. Bks. Bull. 57:483+, 1961.

G57. Tabulae Biologicae. Berlin, Junk, v. 1, 1925-   . 570 T116
   See: Wa p. 145 (note).

G58. Wang, C. H. & Willis, D. L. Radiotracer Methodology in Biological Science. Prentice-Hall, 1965. 574.018 W18r
   See: Science 150:877, 1965.

Information Activities

G59. Bliss, C. I Statistics in Biology. McGraw-Hill, 1967-   .
(3v. projected) 574.018 B61s
   See: Am. Scientist 56:191A-2A, 1968.

G60. Campbell, R. C. Statistics for Biologists. Cambridge Univ. Pr.,
1967. 574.018 C14s
   See: Am. Scientist 56:163A-4A, 1968.

G61. Conference of Biological Editors. Committee on Form and
Style. Style Manual for Biological Journals. 2d ed. Am. Inst. Biol.
Sci., 1964. 029.6 C76s
   See: Wi-EC21.

G62. Fisher, R. A. & Yates, R. Statistical Tables for Biological,
Agricultural and Medical Research. 6th ed., London, Oliver & Boyd,
1963. q311 F536st
   See: Wa p. 145.

G63. Jepson, M. Biological Drawings. 2d ed. London, Murray,
1939-40. 2v. q570 J46b2

G64. Mackay, R. S. Bio-Medical Telemetry: Sensing and Transmitting
Biological Information from Animals and Man. Wiley, 1968. 591.1078
M19b
   See: Science 163:921-2, 1969. Lib. J. 93:2674, 1968.

G65. Mathematical Biosciences. Elsevier, v. 1, 1967-   .
574.19105 MA

G66. Simonton, W. & Mason, C. eds. Information Retrieval with
Special Reference to the Biomedical Sciences. Minneapolis, Nolte
Center, 1966.

G67. Stacy, R. W. & Waxman, B. D. eds. Computers in Biomedical
Research. Academic, 1965. 2v. 574.018 Stlc
   See: Science 150:1576-7, 1965.

G68. Sterling, T. D. & Pollack, S. V. Computers and the Life Sciences.
Columbia Univ. Pr.. 1965. 651.26 St4c
  See: Science 150:1576-7, 1965.

  See also: U. S. National Agricultural Library. Agricultural/Biological
Vocabulary. (No. G32)

Directories

G69. American Institute of Biological Sciences, comp. Directory of
Bioscience Departments in the United States and Canada. Reinhold,
1967. 574.07 D62
  See: Sp. Libs. 59:206, 1968.

G70. Naturalists' Directory. PCL Pub., v. 1, 1878-   500 N219
  See: Wa p. 51.

G71. World Directory of Hydrobiological and Fisheries Institution.
Wash., Am. Inst. Biol Sci., 1963.
  See: Wi-EC111. Wa p. 146.

  See also:
   . Arvey, M. D. & Riemer, W. J. Inland Biological Field Stations
of the U. S. BioScience 16:249-54, 1966.

Serials

G72. U. S. Library of Congress. Biological Sciences Serial Publications; a World List, 1950-1954. Wash., 1955. A570.5 Un3b
  See: Wi-EC5.

G73. U. S. National Library of Medicine. Biomedical Serials,
1950-1960; a Selective List of Serials in the National Library of
Medicine. Wash., 1962. (Public Health Service Pub. No. 910)
016.6105 Un3b
  See: Med. Lib. Assn. Bull. 51:141-2, 1963. Wi-EJ23. Wa p. 186.

BIOCHEMISTRY & BIOPHYSICS

  See also appropriate units in CHEMISTRY, AGRICULTURAL SCIENCES,
    and MEDICAL SCIENCES

G74. Advances in Biological and Medical Physics. Academic, v. 1,
1948-   612.014 Ad95

G75. Altman, P. L. & Dittmer, D. S., eds. Metabolism; Biological Handbook. Wash., Federation of American Societies for Experimental Biology, 1968.

G76. Methods and References in Biochemistry and Biophysics. World Pub., 1966. Companion volume to No. D135.

G77. Progress in Biophysics and Molecular Biology. Pergamon, v. 1, 1950- . (Former title of this annual: Progress in Biophysics and Biophysical Chemistry) 574 P943
  See: Wa p. 150.

  See also:
    Chemical-Biological Activities. (No. D23)
    Bulletin Signalétique. 12: Biophysique. Biochimie. (No. A18)
    Referativnyi Zhurnal. Biologicheskaya khimiya. (No. A19)
    CRC Handbook of Biochemistry with Selected Data for Molecular
      Biology. (No. D133)

BOTANY

Guides to the literature

G78. Lawrence, G. H. M. Literature of Taxonomic Botany. In Taxonomy of Vascular Plants. Macmillan, 1951. p. 284-331. 580. 1 L37t
  See: Wa p. 159 (note).

G79. Lawrence, G. H M. et al. Botanico-Periodicum-Huntianum. Hunt Botanical Library, 1968.
  See: S-H Book News 23:5, (Sept.) 1968.

G80. Stafleu, F. A. Taxonomic Literature; a Selective Guide to Botanical Publications with Dates, Commentaries and Types. Utrect, International Bureau for Plant Taxonomy and Nomenclature, 1967. 016.58014 Stlt
  See also: Ewan, J. Reference Tools for the Botanist. Stechert-Hafner Book News. 20:33-4, (Nov.) 1965.

Bibliographies, Indexes, Abstracts

General

G81. Botanical Abstracts. Williams & Wilkins, v. 1-15, 1918-26. (Continued as Biological Abstracts) 580. 5 BOTA
  See: Wi-EC7. Wa p. 151.

G82. Botanisches Centralblatt. Jena, Fischer. 1880-1919, 1922-45.
580.5 Bs
  See: Wi-EC42. Wa p. 151 (note).

G83. Excerpta Botanica. Stuttgart, Fischer, v. 1, 1959-   . 581.05
EX
  Section A: Taxonomica et chorologica; Section B: Sociologica.
  See: Wi-EC43. Wa p. 151.

G84. Blake, S. F. Geographical Guide to Floras of the World.
Wash., G.P.O., 1942-   . (v. 1, 1942; v. 2, 1960) A581 B58g
  See: Wi-EC33. Wa p. 154.

G85. Pritzel, G.A. Thesaurus Literaturae Botanicae ... Lipsiae,
Brockhaus, 1872-7. Reprinted Milan, Gorlizh, 1950. qA580 P93t
  See: Wi-EC31. Wa p. 151.

G86. Jackson, B.D. Guide to the Literature of Botany. Being a
Classified Selection of Botanical Works, Including Nearly 6000 Titles
Not Given in Pritzel's "Thesaurus". Longmans, 1881. Reprinted
by Hafner, 1964. A580 J13g
  See: Wi-EC30. Wa p. 152.

G87. Torrey Botanical Club. Index to American Botanical Literature.
In Torrey Botanical Club. Bulletin. v. 13, 1886-   . Reprinted and
issued in card form, 1894-   . Complete index, 1886-1966, reprinted
in book form by G.K. Hall, 1968. Supplements will be issued annually
in card form (Hall) with book-form supplement planned every ten years.
  See: Wi-EC44.

G88. U.S. Department of Agriculture. Library. Plant Science Catalog:
Botany Subject Index. G.K. Hall, 1958. 15v. q580 Un3p

G89. American Journal of Botany. Botanical Society of America,
v. 1, 1914-   . 580.5 AMJ

G90. Huntia; a Yearbook of Botanical and Horticultural Bibliography.
Pittsburgh, Hunt Botanical Library. v. 1, 1964-   . 016.58 H92.
  See: Science 152:916-7, 1966. Wa p. 151.

  See also:
    Bulletin Signalétique. 17: Biologie et Physiologie vegetales.
    (No. A18)
    International Catalogue of Scientific Literature: M, Botany.
    (No. A16)

Special

G91. Abstracts of Mycology. Biological Abstracts. v. 1, 1967-   .
589.203 AB

G92. Catalogue of Botanical Books in the Collection of Rachel
McMasters Miller Hunt. Pittsburgh, Hunt Botanical Library. 1958-   .
   See: Wa p. 151 (note).

G93. Harvard University. Arnold Arboretum. Library. Catalogue of
of the Library. Cosmos Pr., 1914-33. 3v. A582 H262c

G94. Index Londinensis to Illustrations of Flowering Plants, Ferns
and Fern Allies, being an Amended and Enlarged Edition, Continued
up to the End of the Year 1920, of Pritzel's "Alphabetical Register of
Flowering Plants and Ferns". Oxford, Clarendon, 1929-31. 6v.
Supplement for 1921-1935. 1941. 2v. q580 In2
   See: Wi-EC36. Wa p. 158.

G95. Langman, I. K. A Selected Guide to the Literature on the
Flowering Plants of Mexico. Univ. of Penn. Pr., 1964. q016.581972
L26s
   See: Science 148:488, 1965.

G96. Massachusetts Horticultural Society. Library. Dictionary
Catalog. G. K. Hall, 1964. 3v.

G97. Nissen, C. Die botanische Buchillustration, ihre Geschichte
und Bibliographie. Stuttgart, Hiersemann, 1951-2. 2v. (Bd. 1,
Geschichte; Bd. 2, Bibliographie) q580 N63b
   See: Wi-EC40.

G98. Sitwell, S. & Blunt, W. Great Flower Books, 1700-1900;
a Bibliographical Record of Two Centuries of Finely Illustrated
Flower Books. Toronto, Collins, 1956.

Histories

G99. Arber, A. Herbals, their Origin and Evolution, a Chapter
in the History of Botany, 1470-1670. 2d ed. Cambridge Univ. Pr.,
1938. 580.9 Arlhl
   See: Wa p. 159 (note).

G100.   Botanical Society of America. Fifty Years of Botany. McGraw-Hill, 1958.   580.4 B65f

G101.   Sachs, J. von. History of Botany, 1530-1860. Oxford, Claredon, 1890. Reprint, 1906. (For continuation see Green No. G99) 590.9 Salg

G102.   Green, J.R. History of Botany, 1860-1900, being a Continuation of Sachs' History. Oxford, Clarendon, 1901. 580.9 G82h

Dictionaries & Encyclopedias

General

G103.   Handbuch der Pflanzenphysiologie. (Encyclopedia of Plant Physiology) Springer, Berlin. 1955-67. 18v. (Contribtions in English and German) 581.1 H191
  See: Wa p. 153.

G104.   Horsfall, J.G. & Diamond, A.E. Plant Pathology; an Advanced Treatise. Academic, 1959- . (5 v. projected) 581.2 H78p
  See: Science 131:1368, 1960; 132:30, 803, 1960.

G105.   Jackson, B.D. Glossary of Botanic Terms with their Derivation and Accent. 4th ed. Lippincott, 1928. Reprinted by Hafner, 1953. 580.3 J12g4
  See: Wi-EC54. Wa p. 152.

G106.   McLean, R.C. & Cook, W.R.I. Textbook of Theoretical Botany. Longmans, 1951- . 580 M222t
  See: Science 162:786, 1968.

G107.   Moldenke, H.N. & Moldenke, **A.L.** **Plants** of the Bible. Chronica Botanica, 1952. 220.858 M73p
  See: Med. Lib. Assn. Bull. 41:304-6, 1953. **Wi-EC70.**

G108.   Snell, W.H. & Dick, E.A. **A Glossary of Mycology.** Harvard Univ. Pr., 1957. 589.203 Sn2g
  See: Science 126:173, 1957. **Wi-EC57.** **Wa p. 160.**

G109.   Steward, F.C. ed. Plant **Physiology.** **Academic,** 1959-
(6v. projected) 581.1 St43p

G110.   Uphof, J.C.T. Dictionary of Economic Plants. 2d ed. Hafner, 1968. 581.603 Up3d.
  See for 1st ed.: Wi-EC70a. Wa p. 154.

G111. Usher, G. A Dictionary of Botany, including Terms used in Biochemistry, Soil Science, and Statistics. Van Nostrand, 1966. 580.3 Us3
   See: Choice 4:408, 1967. Wi-1EC6.

Foreign language

G112. Bedevian, A.K. Illustrated Polyglottic Dictionary of Plant Names in Latin, Arabic, Armenian, English, German, Italian and Turkish Languages. Cairo, Argus & Papazian, 1936. 580.3 B39i
   See: Wi-EC58. Wa p. 159.

G113. Davydov, N.N. comp. Botanical Dictionary: Russian-English-German-French-Latin. 2d ed. Moscow, Fizmatgiz, 1962. 580.3 D31b
   See: Wi-EC59. Wa p. 152.

G114. Gerth van Wijk, H.L. A Dictionary of Plant Names. Hague, Nijhoff, 1911-16. Reprint, 1962. 2v. (v.1: Latin Names; v.2: Index of English, Dutch, French and German names) q580.3 G32d
   See: Wi-EC51.

G115. Steinmetz, E.F. Codex Vegetabilis. 2d ed. Amsterdam, Heinman, 1957. (Names of medicinal plants in Latin, English, European and Oriental languages) 615.32 St3c

G116. Stearn, W.T. Botanical Latin; Grammar, Syntax, Terminology, and Vocabulary. Hafner, 1966. 470 St3b
   See: Wi-1EC5.

Taxonomic literature

G117. International Code of Botanical Nomenclature. Utrecht, International Bureau for Plant Taxonomy and Nomenclature, 1961. 580.1 1n8r
   See: Wi-EC53.

G118. International Code of Nomenclature for Cultivated Plants. Utrecht, International Bureau for Plant Taxonomy and Nomenclature, 1965.

See the bibliography: Bell, C.R. Plant Taxonomy. <u>Bioscience</u>
15:383-5, 1965.

General

G119. Hagerup, O. & Petersson, V. A Botanical Atlas. (Translated by
H. Gilbert-Carter) Copenhagen, Munksgaard, 1959-60. 2v. q581.9489
H12bEg
  See: Wa p. 154.

G120. Harvard University. Gray Herbarium. Gray Herbarium Index.
1873- . Issued quarterly on cards to subscribers. The index
(265,000 cards) was published by G.K. Hall in 10v. in 1968. Updated by quarterly supplements on 1000 cards published by Gray
Herbarium.

G121. Index Kewensis Plantarum ... (an enumeration of the genera
and species of flowering plants from the time of Linnaeus to the year
1885 inclusive, together with their author's name, the works in which
they were first published, their native countries and their synonyms).
1893-5. 2v. Supplements, 1, 1901- . q582 In2
  See: Wi-EC52. Wa p. 158.

G122. International Plant Index. New York Botanical Garden, v. 1,
1962- . 580.14 In8
  See: <u>Science</u> 152:934-5, 1966.

G123. Wit, H.C.D. de. Plants of the World. Dutton, 1966-
(3v. projected) 581 W77p
  See: <u>Lib. J.</u> 92:1022-3, 1967.

Special

G124. Carleton, R.M. Index to Common Names of Herbaceous
Plants. G.K. Hall, 1959. 582.12 C19i
  See: Wi-EC47. Wa p. 313 (note).

G125. Chase, A. comp. Index to Grass Species. G.K. Hall, 1963.
3v. (reproduction of 62,700 cards in the Index) q584.9 C38i
  See: Wa p. 161.

G126. Gleason, H.A. The New Britton and Brown Illustrated Flora of the Northeastern United States and adjacent Canada. N.Y. Botanical Garden, 1952. 3v. 581.97 B77i
  See: Wi-EC73.

G127. Graf, A.B. Exotica 3, Pictorial Cyclopedia of Exotic Plants. Rev. ed. Roehrs Co., 1968. 634.35 G757e

G128. Gray, A. Manual of Botany; a Handbook of Flowering Plants and Ferns of the Central and Northeastern United States and Adjacent Canada. 8th ed. American Book Co., 1950. 581.973 G79m
  See: Science 112:658, 1950. Wi-EC67.

G129. Grimm, W.C. Recognizing Native Shrubs. Stackpole, 1968. 582.17 G88r
  See: Lib. J. 94:560, 1969.

G130. Hutchinson, J. The Genera of the Flowering Plants. Oxford Univ. Pr., v.1, 1964- . 582.13 H97g
  See: Science 147:1561-2, 1965. Wa p. 155 (note).

G131. Jones, G.N. Flora of Illinois; Containing Keys for Identification of Flowering Plants and Ferns. 3d ed. Univ. of Notre Dame, 1963. 581.9773 J71f

G132. King, L.J. Weeds of the World; Biology and Control. Wiley, 1966. 632.7 K584.
  See: Science 92:797, 1967.

G133. Kingsbury, J.M. Poisonous Plants of the United States and Canada. Prentice Hall, 1964. 581.69 K61p
  See: Am. Scientist 52:326A, 1964. Wi-EC68.

G134. Lamb, E. & Lamb, B. The Illustrated Reference on Cacti and other Succulents. London, Blandford, 1955-63. 3v.
  See: Wa p. 315.

G135. Lange, M. & Hora, F.B. Collins' Guide to Mushrooms and Toadstools. London, Collins, 1963. (Translated from the Danish by L. Hansen) 589.222 L26g
  See: Wi-EC76. Wa p. 160.

G136. Mohlenbrock, R.H. The Illustrated Flora of Illinois: Ferns. Southern Ill. Pr., 1967. 587.31 M72

G137. Peterson, R.T. & McKenny, M. A Field Guide to Wildflowers of Northeastern and Northcentral North America. Houghton, 1968. 582.13 P44f
  See: Science News 93:311, 1968.

G138. Rickett, H.W. Wild Flowers of the United States. McGraw-Hill, 1966- . (5v. projected) (v.1, Northeastern; v.2, Southeastern States; v.3, Southwestern; v.4, Pacific Northwest; v.5, Rocky Mountains and the Great Plains) q582.13 R42w
  See: Am. Scientist 55:120A, 178A, 1967; Lib. J. 92:250, 4013, 1967. Science 155:65, 1967.

G139. Sargent, C.S. The Silva of North America. Houghton, 1891-1902. 14v. reprinted by Peter Smith, 1947. 14v. in 7 q582 Sa7
  See: Wi-EC85.

G140. Smith, A.H. The Mushroom Hunter's Field Guide. Univ. of Mich. Pr., 1963. 589.22 Sm6mu
  See: Sci. Am. 212:135, (June) 1964. Wi-EC78.

G141. Viktorov, S.V. et al. Short Guide to Geo-botanical Surveying. Macmillan, 1964. 580 In82 v.8
  See: Science 145:476-7, 1964.

G142. Willis, J.C. A Dictionary of the Flowering Plants and Ferns. 7th ed. Cambridge Univ. Pr., 1966. 580.3 W67d
  See: Science 92:674, 1967; Choice 4:644, 1967.

Biographies & Directories

G143. Barnhart, J.H. comp. Biographical Notes upon Botanists. G.K. Hall, 1965. 3v. q925.8 B26b
  See: Wa p. 153.

G144. Coats, P. Great Gardens of the Western World. Putnam, 1963. (same as Great Gardens. London, Weidenfeld & Nicholson, 1963) q712 C63g
  See: Sub. Bks. Bull. 60:1007-8, 1964.

G145. Lanjouw, J. & Stafleu, F.A. Index Herbariorum; a Guide to the Location and Contents of the World's Public Herbaria. Utrecht, International Bureau for Plant Taxonomy and Nomenclature of the International Association for Plant Taxonomy, 1952- .
  Pt. 1: The Herbaria of the World. 5th ed., 1964; Pt. 2: The Collectors. 1954- . 580.7 L27i
  See: Wi-EC62, 1EC7. Wa p. 153.

G146. Reisigl, H. ed. The World of Flowers. Viking, 1965.
(translation of Blumen-Paradiese der Welt) 634.3 R273bE

G147. Roberts, M.M. Public Gardens and Arboretums of the U.S.
Holt, 1962. 580.744 R54p
　　See: Wi-EC63.

ECOLOGY

　　See: Excerpta Medica, Section XXI: Developmental Biology and
　　　　Teratology. (No. H10)

G148. Advances in Ecological Research. Academic, v.1, 1962-
574.5 Ad9
　　See: Wa p. 150.

G149. Carpenter, J.R. An Ecological Glossary. Univ. of Okla. Pr.,
1938. Reprinted by Hafner, 1956. 575.3 C22e
　　See: Wi-EC49. Wa p. 150.

G150. Hanson, H.C. Dictionary of Ecology. Philosophical, 1962.
574.503 H19d
　　See: Am. Scientist 51:80A, 1963; Science 138:1326, 1962. Wa p. 150.

G151. Gleason, H.A. & Cronquist, A. The Natural Geography of Plants.
Columbia Univ. Pr., 1964. 581.9 G47n
　　See: Science 145: 1172-4, 1964.

G152. Shelford, V.E. The Ecology of North America. Univ. of Ill.
Pr., 1963. 574.5097 Sh4e
　　See: Am. Scientist 52:432A-4A, 1964.

G153. Ecology. Brooklyn, Ecological Society of America, v.1, 1920-   .
570.5 PL

G154. Journal of Animal Ecology. Cambridge Univ. Pr., v.1, 1932-
591.1005 JO

　　See also:
　　Audy, J.R. Emerging Disciplines in the Health Sciences and
　　　　Their Impact on Health Sciences Libraries: Human Ecology.
　　　　Med. Lib. Assn. Bull. 53:410-9, 1965.

117

ENTOMOLOGY

G155. Annual Review of Entomology. Annual Reviews Inc., v.1,
1956-   . 595.7 An78
  See: Wa p. 168.

G156. Chamberlin, W.F. Entomological Nomenclature and Literature.
3d ed. Dubuque, Iowa. W.C. Brown, 1952. 595.7 C35e
  See: Wi-EC122. Wa p. 167 (note).

G157. Horn, W. & Schenkling, S. Index Litteraturae Entomologicae.
Berlin, Dahlem, 1928-29. 4v. (Revision of: Hagen's Bibliotheca
Entomologica, 1862-63. 2v.) A595.7 In2
  See: Wi-EC124. Wa p. 167.

G158. Index to the Literature of American Economic Entomology.
College Park, Md., Entomological Society of America. v.1-18,
1905/14- 1959. (Continues: U.S.D.A. Bibliography of the More
Important Contributions to American Economic Entomology. v.1-18,
1889-1905)

G159. Kéler, S. von. Entomologisches Wörterbuch. 3d ed. Berlin,
Akademie Verlag, 1963. 595.703 K27e
  See: Wa p. 168.

G160. Oppenheimer, J.M. Essays in the History of Embryology and
Biology. M.I.T. Pr., 1967.
  See: Science 157:671, 1967.

G161. Review of Applied Entomology. London, Commonwealth Institute
of Entomology. v.1, 1913-   . (Series A: Agricultural. Series B:
Medical and Veterinary) 595.705 REV
  See: Wi-EC126. Wa p. 167.

  See also:
    Bibliography of Agriculture (No. J8)
    Zoological Record. (No. G193)

GENETICS

G162. Advances in Genetics. Academic. v.1, 1947-
575.1 Ad95

G163. Annual Review of Genetics. Annual Reviews Inc., v.1, 1967-
575.1 An7
  See: Science 159:1091, 1948.

118

G164. Clapper, R.B. Glossary of Genetics and other Biological Terms. Vintage, 1961. 575.103 C53g

G165. King, R.C. A Dictionary of Genetics. Oxford Univ. Pr., 1968. 575.103 K58d

G166. Dunn, L.C. A Short History of Genetics. McGraw-Hill, 1965. 575.109 D92s
    See: Bioscience 17:355, 1967.

G167. Ravin, A.W. The Evolution of Genetics. Academic, 1965. 575.1 R192e
    See: Science Bks. 1:155, (Dec.) 1965.

G168. Sturtevant, A.H. A History of Genetics. Harper & Row, 1965. 575.109 St97h
    See: Science 152:922, 1966.

MICROBIOLOGY

   Bibliographies, Indexes, Abstracts, Reviews

G169. Abstracts of Bacteriology. v.1-9, 1917-26. Merged with Botanical Abstracts to become Biological Abstracts. 589.05 AB

G170. Advances in Applied Microbiology. Academic, v.1, 1959- 589.95 Ad95

G171. Annual Review of Microbiology. Annual Reviews Inc., v.1, 1947- . 589.95 An78
    See: Wa p. 147.

G172. Bacteriological Reviews. Williams & Wilkins, v.1, 1937- . 589.05 BA

G173. Grainger, T.H. A Guide to the History of Bacteriology. Ronald, 1958. 016.58995 G76g
    See: Wi-EC117. Wa p. 147.

G174. Institut Pasteur. Bulletin. Paris, Masson. v.1, 1903- . 570.6 PA
    See: Wi-EC119. Wa p. 147.

G175. Microbiology Abstracts. London, Information Retrieval Inc., v.1, 1965- .

119

See also:

    Bulletin Signalétique. 14: Microbiologie. Immunologie.
     Genetique (No. A18)
    International Catalogue of Scientific Literature: R, Bacteriology.
     (No. A16)

Histories

G176. Bulloch, W. The History of Bacteriology. Oxford, 1938.
589.95 B87h

G177. Doetsch, R.N. ed. Microbiology; Historical Contributions
from 1776-1908. Rutgers Univ. Pr., 1960. 576.082 D67m

G178. Lechevalier, H.A. & Solotorovsky, M. Three Centuries of
Microbiology. McGraw-Hill, 1965. 576.09 L49t

Dictionaries, Encyclopedias, Comprehensive works

G179. Burnet, F.M. The Viruses. Academic, 1959. 3v. 589.95
B934v
  See: Science 131:657, 724 and 929, 1960.

G180. Gibbs, B.M. & Shapton, D.A. Identification Methods for
Microbiologists. Academic, 1968. 576 G35i

G181. Gunsalus, I.C. & Stanier, R.Y. eds. The Bacteria. Academic,
1960-64. 5v. 589.95 G95b
  See: Science 133:750, 1961.

G182. Jacobs, M.B. et al. Dictionary of Microbiology. Van Nostrand,
1957. 589.9503 J15d
  See: Science 126:847, 1957. Wi-EC13. Wa p. 147.

G183. Jacobs, M.B. & Gerstein, M.J. Handbook of Microbiology.
Van Nostrand, 1960. 576 J15h
  See: Science 133:2005, 1961. Wa p. 147.

G184. Topley, W.W.C. Topley and Wilson's Principles of Bacteriology and Immunity. 5th ed. Williams & Wilkins, 1964. 2v. 616.014
T62p

See also section MEDICAL SCIENCES

ZOOLOGY

Guides to the literature

G185. Smith, R.C. & Painter, R.H. Guide to the Literature of the
Zoological Sciences. 7th ed., Burgess, 1966. 016.59 Sm6g
  See: Wi-EC87. Wa p. 162.

G186. Wood, C.A. An Introduction to the Literature of Vertebrate
Zoology. Oxford Univ. Pr., 1931. A596 W85i
  See: Wi-EC94. Wa p. 169.

Bibliographies, Indexes, Abstracts

G187. Bibliotheca Historico-Naturalis (Bibliotheca Zoologica).
Leipzig, Engelmann, 1846. A590 En3b
  See: Wi-EC89.

G188. Bibliotheca Zoologica I. Leipzig, Engelmann, 1861. 2v.
A590 B471
  See: Wi-EC89.

G189. Bibliotheca Zoologica II. Leipzig, Engelmann, 1887-1923.
8v. A590 B471
  See: Wi-EC90.

G190. Harvard University. Museum of Comparative Zoology. Library
Catalogue. G.K. Hall, 1968. 8v. 016.59 H26c

G191. Index-Catalogue of Medical and Veterinary Zoology. G.P.O.
v.1-18, 1932-52. Supplements, 1953- . A614.9 Un31.
  See: Doss, M.A. The Index Catalogue of Medical and Veterinary
Zoology. Med. Lib. Assn. Bull. 41:110-3, 1953. Wi-EJ143.

G192. McGill University. A Dictionary Catalogue of the Blacker-Wood
Library of Zoology and Ornithology. G.K. Hall, 1967.
  See: Lib. J. 92:726, 1967.

G193. Zoological Record. London, Zoological Society of London.
v.1, 1865- . A590 Z75.
  See: Smith's Guide to the Literature of the Zoological Sciences (No.
G182) p. 53-6. Wi-EC99. Wa p. 162.

  See also:
    Bulletin Signalétique. 16, Biologie et physiologie animales.
    (No. A18)

International Catalogue of Scientific Literature: N, Zoology.
  (No. A16)
Bibliography of Agriculture. (No. J8)

Histories

G194. Petit, G. & Theorides, J. Histoire de la Zoologie. Paris,
Herman, 1962. 590.9 P44h

G195. Hall, T.S. Source Book in Animal Biology. McGraw-Hill,
1951. 570 H14s

Dictionaries & Encyclopedias

G196. Florkin, M. & Scheer, B.T. eds. Chemical Zoology.
Academic, 1967-8. 2v. 591.192 F66c

G197. Kaestner, A. Invertebrate Zoology. Wiley, 1967-   4v.
(Translated from 2d ed. of Lehrbuch der speziellen Zoologie.)
590.1 K111E1
  See: Choice 5:1160, 1968; Am. Scientist 56:475A, 1968.

G198. The Larousse Encyclopedia of Animal Life. McGraw-Hill,
1967. q591 L328
  See: Science 158:898-9, 1967. Lib. J. 92:3626-7, 1967.

G199. Leftwich, A.W. A Dictionary of Zoology. 2d ed. Van
Nostrand, 1967. 591.03 L522d
  See: Am. Scientist 52:235A, 1964. Wi-EC103. Wa p. 163.

G200. Pennak, R.W. Collegiate Dictionary of Zoology. Ronald,
1964. 590.3 P38c
  See: Am. Scientist 52:235A, 1964. Choice 1:363, 1964. Wi-EC104.
Wa p. 163 (note).

G201. Traité de Zoologie; Anatomie, Systématique, Biologie.
Paris, Masson, v.1, 1948- . 590 T68
  See: Wi-EC101. Wa p. 162-3.

Taxonomic literature

  See: Smith, R.C. & Painter, R.H. Guide to the Literature of the
    Zoological Sciences. p. 199-220. (No. G185)

G202. International Code of Zoological Nomenclature. London, International Trust for Zoological Nomenclature, 1964. 590.1 In8c
See: Wa p. 165-6.

G203. Blackwelder, R.E. Taxonomy; a Text and Reference Book. Wiley, 1967. 590.12 B568t
See: Science 159:184-5, 1968.

G204. Neave, S.A. Nomenclator Zoologicus; a List of the Names of Genera and Subgenera in Zoology from the 10th edition of Linaeus, 1758, to the end of 1945. London, Zool. Soc. of London, 1939-1950. 5v. (Consult Biological Abstracts or Zoological Record for new names since 1945) 590.1 N27n
See: Wi-EC102. Wa p. 166.

G205. Rothschild. N.M.V. Classification of Living Animals. 2d ed., Longmans, 1965. 59o.12 R74c

G206. Simpson, G.G. Principles of Animal Taxonomy. Oxford Univ. Pr., 1961. 590.12 Si5p

Handbooks, Manuals, Descriptive works

G207. The Animal Kingdom, the Strange and Wonderful Ways of Mammals, Birds, Reptiles, Fishes and Insects. Doubleday, 1954. 3v.
591.5 An54

G208. Blair, W.F. et al. Vertebrates of the United States. 2d ed., McGraw-Hill, 1968. 596 B57v
See: Choice 5:984, 1968.

G209. Collins, H.H., Jr. Complete Field Guide to American Wildlife; East, Central and North. Harper, 1959. 591.97 C69c
See: Sub. Bks. Bull. 56:143+, 1959.

G210. Ennion, E.A.R. & Tinbergen, N. Tracks. Oxford Univ. Pr., 1967. 591.5 En6t
See: Science 160:658-9, 1968.

G211. Hyman, L.H. The Invertebrates. McGraw-Hill, 1940-59. 5v.
593 H991i
See: Science 96:219, 1940. Wa p. 166.

G212. Moore, C.B. Book of Wild Pets. Putnam, 1937. Reprint, Boston, Branford. 1954. 591.5 M78b
See: Booklist 34:172, 1938.

G213. National Geographic Society. Wild Animals of North America. Wash., 1960. 599 N21w
  See: Sub. Bks. Bull. 57:201+, 1960.

G214. Rue, L. L. Sportsman's Guide to Game Animals; a Field Book of North American Species. Harper, 1968.
  See: Booklist 65:387, 1968.

Directories

G215. International Union of Biological Sciences. Index des Zoologistes. Paris, 1953. Supplement, 1959. 590 In8i
  See: Wa p. 164.

G216. International Zoo Yearbook. London, Zoological Society of London. v. 1, 1959- . (Annual) 590.744 In8
  See: Science 162:659-60, 1968.

G217. Kirchshofer, R., ed. The World of Zoos: a Survey and Gazetteer. Viking, 1968. 590.744 K63zEm
  See: RQ 8:142, 1969.

Birds

Bibliographies

G218. Sitwell, S. et al. Fine Bird Books, 1700-1900. London, Collins, 1953. fA598.2 Si8f

G219. Strong, R. M. Bibliography of Birds. Chicago, Field Museum of Natural History, 1939-59. 4v. 590.5 FI v.25
  See: Wi-EC107. Wa p. 170 (note).

See bibliography in Bioscience 18:62-5, (Jan.) 1968.

Handbooks, Checklists, etc.

G220. Alexander, W. B. Birds of the Ocean: Containing Descriptions of all of the Sea-Birds of the World, with Notes on their Habits and Guides to their Identification. 2d ed. Putnam, 1954. Reprint, 1963.

G221. Bent, A.C. Life Histories of North American Birds. U.S. National Museum, 1919-58. 20v. (originally issued as bulletins) 506 Un35b
Abridgment issued under editorship of H.H. Collins Jr. in 2v., 1960.

G222. Fisher, J. & Peterson, R.T. The World of Birds. Doubleday, 1964. q598.2 F534w
See: Sat. Rev. Mar. 20, 1965 p. 51. Sci. Am. 212:135-6, (Mar.) 1965.

G223. Gilliard, E.T. Living Birds of the World. Doubleday, 1958. 598.2 G411
See: Sub. Bks. Bull. 55:145+, 1958.

G224. Grossman, M.L. & Hamlet, J. Birds of Prey of the World. Potter, 1964. q598.9 G91b
See: Sat. Rev. Mar. 20, 1965 p. 51.

G225. Hanzak, J. The Pictorial Encyclopedia of Birds. Crown, 1968. 598.2 H58p

G226. Peters, J.L. Checklist of Birds of the World. Cambridge, Mass., Museum of Comparative Zoology. v.1, 1931- . 598.2 P44c

G227. Robbins, C.S. et al. Birds of North America. Golden, 1966. 598.2973 R53b
See: Audubon Magazine 60:376, 1966.

G228. Thomson, A.L., ed. A New Dictionary of Birds. **McGraw-Hill,** 1964. 598.203 T38n
See: Sub. Bks. Bull. 62:174-5, 1966. Wi-1EC9. Wa p. 171.

G229. Wetmore, A. Song and Garden Birds of North America. **1964.** (Portraits and text correlated with an album of reproductions of songs) 598.2973 W53s

Fishes

See bibliography in BioScience 15:49-50, 1965.

G230. Axelrod, H.R. & Schultz, L.P. Handbook of Tropical Aquarium Fishes. McGraw-Hill, 1955. 597.5 Ax3h
See: Wi-EC108.

G231. Gabrielson. I.N., ed. The Fisherman's Encyclopedia. Rev. ed., Stackpole, 1963. q799.1 F539
See: Lib. J. 89:648, 1964.

G232. Herald, E.S. Living Fishes of the World. Doubleday, 1961.
597 H411
   See: Lib. J. 86:1780, 1961. Wi-EC109.

G233. Nieuwenhuizen, A. van den. Tropical Aquarium Fish. Van
Nostrand, 1964. 597.5 N55eEl
   See: Lib. J. 89:4924, 1964.

G234. Sterba, G. Freshwater Fishes of the World. London, Vista
Books, 1962. 574.92074 St4sEt
   See: Wa p. 170.

G235. Walden, H.T. II. Familiar Freshwater Fishes of America.
Harper, 1964.
   See: Lib. J. 89:3174, 1964.

Mammals

   See bibliography in Bioscience 16:291-2, 1966.

G236. Burt, W.H. A Field Guide to the Mammals: Field Marks of all
Species North of the Mexican Boundary. 2d ed. Houghton, 1964.
599 B95f
   See: Lib. J. 89:4509, 1964.

G237. Burton, M. Systematic Dictionary of Mammals of the World.
London, Museum Pr. (N.Y., Crowell), 1962. 599.012 B95s
   See: Wi-EC112. Wa p. 172.

G238. Hall, E.R. & Kelson, K.R. The Mammals of North America.
Ronald, 1959. 2v. 599 H14ma
   See: Sub. Bks. Bull. 56:47, 1959. Wi-EC113.

G239. Hoffmeister, D.F. & Mohr, C.O. Fieldbook of Illinois Mammals.
Urbana, Ill., 1957. (Illinois Natural History Survey Manual No. 4)
599 H673f

G240. Ingles, L.G. Mammals of the Pacific States: California,
Oregon, and Washington. Stanford Univ. Pr., 1965. 599 In45ma
   See: Science 150:732-3, 1965.

G241. Napier, J.R. & Napier, P.H. A Handbook of Living Primates.
Academic, 1967. 599.8 N16h
   See: Am. Scientist 56:270A-1A, 1968.

G242. Morris, D. The Mammals: a Guide to the Living Species.
Harper, 1965. 599 M83m
   See: Wi-1EC10.

G243. Palmer, E.L. Fieldbook of Mammals. Dutton, 1957. 599
P184f
   See: Sub. Bks. Bull. 54:298+, 1957.

G244. Walker, E.P. et al. Mammals of the World. 2d ed., Johns
Hopkins Pr., 1968. 2v. 599 W151m
   For review of 1st ed. see: Science 146:1285-6, 1964. Wi-EC114.

   Reptiles & Amphibians

      See bibliography in Bioscience 15:442, 1965.

G245. Bishop, S.C. Handbook of Salamanders: the Salamanders of the
United States, of Canada, and of Lower California. Comstock, 1943.
Reprint, Hafner, 1962; Cornell Univ. Pr., 1967. 597.9 B54h
   See: Lib. J. 92:1635, 1967; Choice 4:635, 1967.

G246. Peters, J.A. Dictionary of Herpetology. Hafner, 1964.
598.103 P44d
   See: Science 147:43, 1965. Wa p. 170 (note).

G247. Stebbins, R.C. A Field Guide to Western Reptiles and
Amphibians; Field Marks of all Species in Western North America.
Houghton, 1966. 598.1 St31f
   See: Lib. J. 92:997, 1967.

G248. Wright, A.H. & Wright, A.A. Handbook of Snakes of the
U.S. & Canada. Comstock, 1957. 2v. 598.12 W92h
   See: Science 126:216, 1957.

G249. Zim, H.S. & Smith, H.M. Reptiles and Amphibians. Simon &
Schuster, 1953. 598.1 Z65r

CURRENT INFORMATION; SELECTION AIDS

   Current issues of such periodicals as:
      American Biology Teacher, v.1, 1938-
      American Naturalist, v.1, 1867- .
      Audubon Magazine, v.1, 1899- .
      Bioscience, v.1, 1951- .

Human Biology, v. 1, 1929-   .
National Geographic, v. 1, 1899-
Natural History, v. 1, 1900-
Quarterly Review of Biology, v. 1, 1926-
Stain Technology, v. 1, 1925-

The medical sciences are concerned with the prevention, cure, and alleviation of disease in man.   This group of sciences may be categorized as:

Basic medical sciences such as anatomy, physiology, biochemistry, microbiology, pathology, pharmacology.

Clinical medical sciences such as internal medicine, surgery, neurology, dermatology, obstetrics, pediatrics, urology, etc.

Allied medical sciences such as dentistry, pharmacy, nursing, etc.

GUIDES TO THE LITERATURE

H1. Blake, J. B. & Roos, C., eds. Medical Reference Works, 1679-1966 a Selected Bibliography. Medical Library Assn., 1967. (Med. Lib. Assn. Pub. No. 3) Supersedes the bibliographies published as part of the Medical Library Association's Handbook of Medical Library Practice. 2d ed., 1956.   027.9 M4891p no.3
  See: **Med. Lib. Assn. Bull.** 56:345-6, 1968. **Lib. J.** 93:2830-1, 1968.

H2. Medical Library Association. Handbook of Medical Library Practice. 3d ed. Chicago. Announced for 1969.

H3. Morton, L. T. How to Use a Medical Library; a Guide for Practitioners, **Research Workers and Students.** 4th ed. Heineman, 1964. 027.9 M84h
  See: Wa **p.** 180.

H4. North Carolina. University. School of Library Science. An Introduction to the Literature of the Medical Sciences. 2d ed., Univ. of N.C. Book Exchange, 1967. 016.61 N81i

H5. U.S. Public Health Service. Film Reference Guide for Medicine and Sciences. Wash., 1956-   . (P.H.S. Pub. 487 rev.) 017 F487

  See also: Brandon, A.N. Selected List of Books and Journals for the Small Medical Library. **Med. Lib. Assn. Bull.** 57:130-50, 1969.

BIBLIOGRAPHIES, INDEXES, ABSTRACTS

General

See also: U.S. National Library of Medicine. Index-catalogue
of the Library ... (No. A17)

H6. Abstracts of World Medicine. London, British Medical Assn.,
v.1, 1947-   . 610.5 AB
See: Wi-EJ38. Wa p. 177.

H7. Biographisches Lexikon derhervorragenden Ärzte aller Zeiten und
Völker. Berlin, Urban, 1929-35. 5v. Reprinted 1962. 926.1 B52
See: Wi-EJ85.

H8. Biographisches Lexikon der hervorragenden Ärzte der letzten
50 Jahre. Berlin, Urban, 1932-33. 2v. (continues No. H7) 926.51
B52
See: Wi-EJ86.

H9. Current List of Medical Literature. Wash., Army Medical Library,
v.1-36, 1941-59. Reprinted by Johnson Reprint. (Succeeded by Index
Medicus new series. No. H11) 610.5 CU
See: Wi-EJ29. Wa p. 179 (note).

H10. Excerpta Medica; the International Medical Abstracting Service.
Amsterdam, Excerpta Medica. v.1, 1947-   610.5 EXC
See: Wi-EJ39 and IEJ6. Wa p. 178.
Issued in the following sections:

Section 1.   Anatomy, Anthropology, Embryology and Histology, v.1,
                   1947-   .
         2A.   Physiology v. 18, 1865-   . v.1-17, 1948-64 in combination
                   with sections 2B and 2C.
         2B.   Biochemistry. v.18, 1965-   (see 2A)
         2C.   Pharmacology and Toxicology. v.18, 1965-   (see 2A)
         3.   Endocrinology. v.1, 1947-
         4.   Medical Microbiology and Hygiene. v.1-7, 1945-54.
         4A.   Medical Microbiology, Immunology and Serology. v.8, 1955-
         5.   General Pathology and Pathological Anatomy. v.1, 1948-
         6.   Internal Medicine. v.1, 1947-   .
         7.   Pediatrics. v.1, 1947-   .
         8A.   Neurology. v.1, 1948-   . (combined with 2B, 1948-65)
         8B.   Psychiatry. v.1, 1948-   . (see 8A)
         9.   Surgery. v.1, 1947-

9B.   Orthopedic surgery. v.1, 1956-   . v.1-10, 1956-65 has title Orthopedics and Traumatology.
10.   Obstetrics and Gynecology. v.1, 1948-   .
11.   Otolaryngology and Rhinolaryngology. v.1, 1948-   .
12.   Ophthalmology. v.1, 1947-   .
13.   Dermatology and Venereology. v.1, 1947-   .
14.   Radiology. v.1, 1947-   .
15.   Chest Diseases. v.1, 1948-   . (Title varies)
16.   Cancer. v.1, 1953-   .
17.   Public Health, Social Medicine and Hygine. v.1, 1955-   . 1957-   .
18.   Cardiovascular Diseases and Cardiovascular surgery. v.1, 1957-   .
19.   Rehabilitation. v.1, 1958-   .
20.   Gerontology and Geriatrics. v.1, 1958-   .
21.   Developmental Biology and Teratology. v.1, 1960-   .
22.   Human Genetics. v.1, 1962-   .
23.   Nuclear Medicine. v.1, 1964-   .
24.   Anesthesiology. v.1, 1966-   .
25.   Hematology. v.1, 1966-   .
26.   Immunology. v.1, 1966-   .
27.   Medical instrumentation. v.1, 1966-   .
28.   Urology. v.1, 1966-   .

H11. New York Academy of Medicine. Author and Subject Catalogs of the Library of the New York Academy of Medicine. G.K. Hall, announced for 1969. Author catalog, 43v.; Subject catalog, 34 v.

H12. Index Medicus. Wash., National Library of Medicine, new series, v.1, 1960-. Monthly.   Cumulates annually into Cumulative Index Medicus.   Available also on magnetic tape. 610.5 CUM
    This title existed 1879-1927 as a "quarterly classified record of current medical literature of the world". Series 1: v.1-21, 1879-1899; series 2: v.1-18, 1903-20; series 3: v.1-6, 1921-27; war supplement, 1914-17. In 1927 it merged with Quarterly Cumulative Index to Current Medical Literature to form Quarterly Cumulative Index Medicus.
    Quarterly Cumulative Index Medicus (QCIM), v.1-60, 1927-56, was published quarterly by the American Medical Association through v.44, 1948, with semiannual cumulations. It became semiannual with v.45, 1949.

    The January issue of the volumes in the new series contains a list of journals indexed, which is supplemented in the monthly issues. This list is also available as a separate title (No. H12).

    Annually Part 2 of the January issue is titled Medical Subject Headings: Main Headings and Cross References Used in Index Medicus and National Library of Medicine Catalog (MESH)

See: Med. Lib. Assn. Bull. 54:325-36, 1966; 55:259-78, 1967.
Wi-EJ26 to EJ32. Wa p. 179.

H13. U.S National Library of Medicine. List of Journals Indexed
in Index Medicus. Wash., 1963- . 610.5 CUM

H14. U.S. National Library of Medicine. Catalog. Wash., Library of
Congress, 1948-1965. Annual with quinquennial cumulations, 1950/54-
1966. (Title varies) Continued by No. H15. 016.61 Un32n
  See: Wi-EJ14.

H15. U.S. National Library of Medicine. National Library of Medicine
Current Catalog. Wash., G.P.O., v.1, 1966- . Biweekly, cumulated
quarterly; annually cumulated into bound volume. 016.61 Un32n.
  See: Med. Lib. Assn. Bull. 57:36-40, 1969; Am. Doc. 18:189-90,
1967. Wi-1EJ2.

H16. U.S. National Library of Medicine. A Catalogue of Sixteenth
Century Printed Books in the National Library of Medicine. Wash.,
G.P.O., 1967. 016.094 Un3c
  See: Med. Lib. Assn. Bull. 55:537, 1967.

H17. Current Medical References. Lange Medical Pubs., v.1, 1959- .
(revised frequently; editors vary. 5th ed., 1967, edited by M. J.
Chatton) 016.61 Sa5c
  See: Wi-1EJ1. Wa p. 196.

H18. Meditsinskii Referativnyi Zhurnal. Moscow, Medgiz, v.1,
1960- . Published in 13 sections. 610.5 MDT
  See: Medical Reference Works p.3-4, Wi-EJ42. Wa p. 180.

H19. Wellcome Historical Medical Library, London. A Catalogue of
Printed Books in the Wellcome Historical Medical Library. London,
v.1, 1962- . q016.61 W45c
  See: Wi-EJ19; 1EJ3. Wa p. 181.

H20. American Medical Association. Journal. Chicago, v.1,
1883- . 610.5 AM

H21. Medical Library Association. Bulletin. Chicago, v.1, 1911- .
027.905 ME

    See also:
      Biological Abstracts (No. G7)
      Bulletin Signalétique. Sections 12-18. (No. A18)
      Chemical Abstracts (No. D14)
      International Abstracts of Biological Sciences (No. G12)
      Referativnyi Zhurnal (No. A19)
      Reuss. Repertorium Commentationum ... v.10-16. (No. A14)

Special

H22. Aerospace Medicine and Biology, a Continuing Bibliography.
Wash., Library of Congress, v. 1, 1952- . (title varies) 016.61698
Ae82
  See: Wi-EJ44. Wa p. 197.

H23. Guerra, F. American Medical Bibliography, 1639-1783.
T. C. Harper, 1962. 016.610973 G93a
  See: Wi-EJ78. Wa p. 181.

H24. Dental Abstracts. Chicago, American Dental Assn., v. 1, 1956-
  See: Wi-EJ102. Wa p. 199.

H25. Index to Dental Literature. Chicago, American Dental
Assn., v. 1, 1921- . (title, coverage, arrangement, and frequency
vary)
  See: Medical Reference Works p. 135. Wi-EJ100, 1EJ18.

H26. International Pharmaceutical Abstracts. Wash., American
Society of Hospital Pharmacists. v. 1, 1964-
  See: Wa p. 193.

H27. RINGDOC. Derwent Pooled Pharmaceutical Literature Documentation: Abstracts Journal. London, v. 1, 1964-

H28. Unlisted Drugs. Special Libraries Assn., v. 1, 1949- .
  See: Wi-EJ134.

H29. U. S. National Library of Medicine. Toxicity Bibliography.
Wash., v. 1, 1968-

H30. Cumulative Index to Nursing Literature. Glendale, Calif.,
Seventh-Day Adventist Hospital Assn., v. 1, 1956- . (quarterly
with annual cumulations) 016.61073 C91
  See: Wi-EJ109.

H31. International Nursing Index. American Journal of Nursing Co.,
v. 1, 1966- . (Quarterly. Annual cumulation)
  See: Med. Lib. Assn. Bull. 54:286-7, 1966; Sp. Libs. 59:368, 1968.

H32. Hospital Literature Index. Chicago, American Hospital Assn.,
v. 1, 1945- . (title varies)
  Quarterly, with annual cumulation. Quinquenially cumulated as
Cumulative Index of Hospital Literature.
  See: Wi-EJ35.

REVIEWS & SURVEYS

H33. Advances in Biomedical Engineering and Medical Physics. Wiley, v. 1, 1968- . 610.78 Ad9

H34. Advances in Internal Medicine. Year Book Medical Pub., v. 1, 1942- . (Biennial) 616 Ad95

H35. Annual Review of Medicine. Annual Reviews, v. 1, 1950- . 610 An78

H36. Medical Progress: a Review of Medical Advances. N. Y., v. 1, 1952- .

H37. Methods in Medical Research. Year Book Medical Pub., v. 1, 1948- . 610.7 M56

H38. Physiological Reviews. Baltimore, American Physiological Society. v. 1, 1921- . (Quarterly) 612.05 PHR

H39. U.S. National Library of Medicine. Monthly Bibliography of Medical Reviews. Wash., G. P. O., v. 1, 1968- . Supersedes Bibliography of Medical Reviews, v. 1-12, 1955-67.
  Also included in monthly issues of Index Medicus and cumulated in Cumulated Index Medicus.
  See: Wi-EJ36.

H40. Year Book of Medicine. Year Book Medical Pub., v. 1, 1901- . (Title varies) 610 Y319

HISTORIES

  See:
    Blake & Roos., eds. Medical Reference Books. p. 50-91. (No. H1)
    Early History of Medicine. Choice 6:187-91, 1969.

H41. Bibliography of the History of Medicine. National Library of Medicine, v. 1, 1965- . 016.6109 Un32b
  See: Sp. Libs. 57:727, 1966.

H42. Miller, G. ed. Bibliography of the History of Medicine of the United States and Canada, 1939-1960. Johns Hopkins Pr., 1964. 016.61 M56b
  See: Med. Lib. Assn. Bull. 53:489-91, 1965. Wi-EJ75.

H43. Castiglioni, A. A History of Medicine. 2d ed. Knopf, 1958.
610.9 C27s
  See: Wa p. 187.

H44. Garrison, F.H   An Introduction to the History of Medicine.
4th ed. Saunders, 1929. Reprinted 1960. 610.9 G19i
  See: Wa p. 187.

H45. Garrison, F.H. & Morton, L.T. Garrison and Morton's Medical
Bibliography, an Annotated Checklist of Texts Illustrating the History
of Medicine. 2d ed. London, Grafton; 1954. Reprinted, 1965.
A610 G19m
  See: Wi-EJ77. Wa p. 188.

H46. Kett, J.F. The Formation of the American Medical Profession:
The Role of Institutions, 1780-1860. Yale Univ. Pr., 1968.
  See: Am. Scientist 56:499A-500A, 1968.

H47. Major, R.H. A History of Medicine. Thomas, 1954. 2v.
610.9 M28h
  See: Med. Lib. Assn. Bull. 43:329-30, 1955.

H48. Mettler, C.C. History of Medicine; a Correlative Text, Arranged
According to Subjects. Blakiston, 1947. 510.9 M56

H49. Sigerist, H.E. History of Medicine. Oxford Univ. Pr., 1951-61.
2v. 610.9 Si2h
  See: Wa p. 188.

H50. Singer, C. & Underwood, E.A. A Short History of Medicine.
2d ed. Oxford, Clarendon, 1962.
  See: Wa p. 188.

H51. Thornton, J.L. Medical Books, Libraries, and Collectors: a
Study of Bibliography and the Book Trade in Relation to the Medical
Sciences. 2d ed. London, Deutsch, 1966. 610 T39m
  See: Med. Lib. Assn. Bull. 55:252-3, 1967.

DICTIONARIES, ENCYCLOPEDIAS, TERMINOLOGIES

  General

H52. American Medical Association. Current Medical Terminology.
3d ed., Chicago, 1966. 610.3 C93.

H53. Dorland's Illustrated Medical Dictionary. 24th ed., Saunders, 1965. 610.3 D73a
  Abridged edition: Dorland's Pocket Medical Dictionary. 21st ed. Saunders, 1968.
   See: Wi-EJ47; 1EJ8. Wa p. 183.

H54. Harned, J.M. Medical Terminology Made Easy. 2d ed., Physicians' Record, 1968. 610.3 H22m

H55. Parr's Concise Medical Encyclopaedia. Elsevier, 1965.
   See: Wi-1EJ9.

H56. National Conference on Medical Nomenclature. Standard Nomenclature of Diseases and Operations. 5th ed., Blakiston, 1961. 616.01 N21s
   See: Wi-EJ57.

H57. Skinner, H.A. The Origin of Medical Terms. 2d ed. Williams & Wilkins, 1961. 610.3 Sk3o
   See: Wa p. 177.

H58. Stedman, T.L. Stedman's Medical Dictionary. 21st ed., Williams & Wilkins, 1966. 610.3 St3p
   See: Wi-1EJ11. Wa p. 184.

H59. Steen, E.B. Dictionary of Abbreviations in Medicine and Related Sciences. 2d ed., London, Cassells, 1963. 610.148 St3d
   See: Wa p. 177.

H60. Strand, H.R. **An Illustrated Guide to Medical** Terminology. **Williams & Wilkins, 1968. 610.14 St8i**

H61. Strauss, **M.B., ed. Familiar Medical Quotations.** Little, Brown, 1968. **610 St8f**
   See: **Med. Lib. Assn. Bull.** 57:91, 1969.

H62. Taber, C.W. **Cyclopedic Medical Dictionary.** 10th ed., Davis, 1965. 610.3 T11d
   See: Wi-1EJ12. Wa p. **184.**

   Foreign language

H63. De Vries, L. German-English Medical Dictionary. McGraw-Hill, 1952. 610.3 D49g
   See: Wa p. 184.

H64. Lejeune, F. & Bunjes, W.E. Deutsch-Englisches, Englisches-Deutsches Wörterbuch für Ärzte. Stuttgart, Thieme, v.1, German-English, 2d ed, 1968. v.2, English-German, 1953 (reprint, 1968).
　See: Wa p. 184-5.

H65. Lepine, P. Dictionnaire Francais-Anglais, Anglais-Français des Terms médicaux et biologique. Paris, Flammarion, 1952.
610.3 L55d
　See: Wa p. 185.

H66. Jablonski, S. Russian-English Medical Dictionary. Academic, 1958. 610.3 J11r
　See: Wa p. 185.

H67. Elsevier's Medical Dictionary in Five Languages. Elsevier, 1964. (English/American, French, Italian, Spanish, German)
　See: Wi-EJ52.

INFORMATION ACTIVITIES

　See also BIOLOGICAL SCIENCES, unit No. G59-G64

H68. California. University. University at Los Angeles. BMD: Biomedical Computer Programs. 2d ed., Univ. of Calif. Pr., 1967.
510.84 C128p

H69. Herner and Co. A Recommended Design for the United States Medical Library and Information System. Clearinghouse, 1966 2v. (PB 172-923, PB 172-924) 021.63 H558r

H70. Turner, A.H., Jr. & Schmidt, D.A., eds. Computers in Medicine: Bibliography. School of Medicine, Univ. of Mo., 1966.

H71. U.S. National Library of Medicine. MEDLARS: 1963-1967. Wash., G.P.O., 1968. (P.H.S. Pub. No. 1823)
　An updated version of The MEDLARS Story, 1963.

H72. U.S. National Library of Medicine. The National Library of Medicine Classification: a Scheme for the Shelf Arrangement of Books in the Field of Medicine and its Related Sciences. 3d ed., Wash., 1964. 025.4161 Un3c

INFORMATION FOR THE LAYMAN

H73. American Medical Association. Today's Health Guide. Chicago, 1965. 613 R32t

H74. Better Homes and Gardens Family Medical Guide. Meredith, 1964. 616.02 C77b
  See: Sub. Bks. Bull. 61:929-31, 1965.

H75. Field, M. Patients Are People. 3d ed., Columbia, 1967. 362.1 F45p
  See: Lib. J. 92:4169, 1967.

H76. Fishbein, M. Modern Home Remedies and How to Use Them. Doubleday, 1966.
  See: Lib. J. 91:5417, 1966.

H77. Henderson, J. Emergency Medical Guide. 2d ed., McGraw-Hill, 1969.
  See: Science News 95:436, 1969.

H78. Miller, B. The Modern Medical Encyclopedia. Golden, 1965. 12v.

H79. Rossman, I. J. & Schwartz, D. R. The Family Handbook of Home Nursing and Medical care. Lippincott, 1968. 649.8 R73f

H80. Roueche, B. Field Guide to Disease: A Handbook for World Travelers. Little, Brown, 1967. 613.1 R75f
  See: Lib. J. 91:1633-4, 1967.

H81. Shepard, K.S. Care of the Well Baby: Medical Management of the Child from Birth to 2 Years of Age. Lippincott, 1968.

H82. "Take As Directed"; Our Modern Medicines. Chemical Rubber, 1967. 615 T139
  See: Lib. J. 94:1584, 1969.

H83. Today's Health. Chicago, American Medical Assn., v. 1, 1923- Title varies. 610.5 AY

TREATISES, COMPREHENSIVE WORKS

Anatomy

H84. Blinkov, S.M. & Glezer, I.I. The Human Brain in Figures and Tables; a Quantitative Handbook. (Tr. from the 1964 Russian edition) Basic Books, 1968. 611.81 B61mEh
  See: Science 161:454-5, 1968.

H85. Cunningham, D. J. Manual of Practical Anatomy. 13th ed.,
Oxford Univ. Pr., v. 1, 1968- . 3v. 611 C91m
  Concise treatment in: Cunningham's Textbook of Anatomy. 10th ed.,
Oxford Univ. Pr., 1964. 611 C91t

H86. Gray, H. Anatomy of the Human Body. 28th ed., Lea & Febiger,
1966. 611 G79a
  See: Wa p. 190.

Medical Biochemistry & Physiology

See also CHEMISTRY and BIOLOGICAL SCIENCES

H87. American Physiological Society. Handbook of Physiology.
Williams & Wilkins, v. 1, 1959- . q612 Am3h

H88. Beaton, G. & McHenry, E. W., eds. Nutrition: a Comprehensive
Treatise. Academic Pr., v. 1, 1964- . 641.1 B38n

H89. Oser, B. Hawk's Physiological Chemistry. 14th ed., Blakiston,
1965. 612.01 H31p

Microbiology & Pathology

See section BIOLOGICAL SCIENCES

H90. Boyd, W. C. Fundamentals of Immunology. 4th ed. Interscience,
1966. 615.36 B69f

H91. Boyd, W. C. Pathology for the Physician. 8th ed., Lea & Febiger,
1967.

H92. Davis, B. D. et al. Microbiology. Harper & Row, 1967. 576
M583
  See: Science 160:979, 1968.

H93. Gibbs, B. M. Identification Methods for Microbiologists.
Academic, 1966. 576 G35i

H94. Society of American Bacteriologists. Bergey's Manual of
Determinative Bacteriology. 7th ed. Williams & Wilkins, 1957.
589.95 So13b
  See: W1-EC120.

H95. Index Bergeyana; an Annotated Alphabetic Listing of Names of the Taxa of the Bacteria. Williams & Wilkins, 1966. 589.9 In2
  A companion volume to No. H94.
  See: Wi-1EC11.

H96. Zinsser, H. Microbiology. 14th ed. Appleton, 1968. 589.95 H62t

Pharmacology, Pharmacy, Toxicology

General

H97. Chatten, **L.G. ed. Pharmaceutical Chemistry.** Dekker, 1966. 2v.
  See: New Tech. **Bks.** 51:354, 1966.

H98. Dreisbach, R.H. Handbook of Poisoning, Diagnosis and Treatment. 5th ed. Lange Medical Pubs., 1966. 615.9 D81h

H99. Gleason, M.N. et al. Clinical Toxicology of Commercial Products; Acute Poisoning (Home & Farm). 3d ed. Williams & Wilkins, 1969. 615.9 G47c

H100. Goodman, L.S. & Gilman, A. The Pharmacological Basis of Therapeutics. 3d ed. Macmillan, 1965. 615 G62p

H101. Grollman, A. Pharmacology and Therapeutics. 6th ed., Lea & Febiger, 1965. 615.7 G89p

H102. The Merck Manual of Diagnosis and Therapy. 11th ed., Merck & Co., 1966. (Title varies) 615 M53

H103. Plunkett, E.R. Handbook of Industrial Toxicology. Chemical Pub., 1966. 615.9 P74h
  See: Wi-1EJ14.

H104. Spector, W.S. ed. Handbook of Toxicology. Saunders, 1956-59. 5v. (v.1: Acute toxicities; v.2: Antibiotics; v.3: Insecticides; v.4: Tranquilizers; v.5: Fungicides) 615.9 Sp3h
  See: Wa p. 196.

H105. American Medical Association. New Drugs. Chicago, v. 1,
1965-   (annual) Replaces New and Non-official Drugs, 1909-64.
615.1 Am25n

H106. American Pharmaceutical Association. The National
Formulary. 12th ed. Lippincott, 1965. 615.4 Am3n
  See: Wi-EJ128.

H107. Marler, E. E. J. Pharmacological and Chemical Synonyms;
a Collection of more than 13,000 Names of Drugs and Other Compounds Drawn from the Medical Literature of the World. 3d ed.,
Amsterdam, Excerpta Medica, 1961. 615.103 M34p

H108. Modern Drug Encyclopedia and Therapeutics Index. 10th ed.,
R. H. Donnelly, 1965. (Supplemented between revisions by Modern
Drugs, 1941-65) 615 M72

H109. Pharmacopoeia of the United States of America. (USP XVII).
17th ed. Mack Printing Co., 1965. Supplements, 1966-   . (Revised every 5 years) 615.11 P4941
  See: Wi-EJ132.

H110. Physicians' Desk Reference to Pharmaceutical Specialties
and Biologicals (PDR). Rutherford, N. J. Medical Economics,
1947-   . (Annual with supplements) 615 P569

H111. Remington, J. P. The Practice of Pharmacy; a Treatise on
Making, Standardizing, and Dispensing. 12th ed. Lippincott, 1961.
615.4 R28p

H112. U. S. National Library of Medicine. Russian Drug Index.
2d ed. Wash., 1967. 615.1014 J11r

H113. U. S. Dispensatory and Physicians' Pharmacology. 26th ed.
Lippincott, 1967. (title varies) 015.12 D63
  See: Wi-EJ129.

  See also:
    Merck Index (No. D140)
    Unlisted Drugs (No. H28)

General Medicine

H114. American Public Health Association. Control of Communicable Diseases in Man. 10th ed., N.Y., 1965. 614.5 Am3c

H115. Beeson, P.B. & McDermott, W. Cecil-Loeb Textbook of Medicine. 12th ed., Saunders, 1967. 1v. or 2v. 616 C32t
  See: Choice 5:1468, 1969.

H116. Bauer, J.D. et al. Bray's Clinical Laboratory Methods. 7th ed., 1968. 616.076 B73s

H117. Davidsohn, I. & Henry, J.B. Todd-Sanford Clinical Diagnosis by Laboratory Methods. 14th ed., Saunders, 1969. 616.07 T56c

H118. Grant, M. Handbook of Preventive Medicine and Public Health. Lea & Febiger, 1967. 614 C76h

H119. Handbook of Clinical Laboratory Data. 2d ed., Chemical Rubber Co., 1968. 616.02 H191

H120. Robinson, G.A. Hospital Administration. 2d ed., London, Butterworth, 1966.

H121. Rosenau, M.J. Maxcy-Rosenau Preventive Medicine and Public Health. 9th ed., Appleton, 1965. 613 R72p

H122. Year Book of _____ Series (18 subjects in field of medical sciences) Year Book Med. Pub., 1901- .

MEDICAL STATISTICS

H123. American Medical Association. Distribution of Physicians, Hospitals, and Hospital Beds in the U.S.: Regional, State, County, Metropolitan Area. Chicago. 1966. 362.11 T34d

H124. National Health Education Committee. Facts on the Major Killing and Crippling Diseases in the U.S. Today. Wash., 1955- . 616 N21f

H125. U.S. National Institutes of Health. Public Health Service
Grants and Awards. Part 1: Research; Part 2: Training. Wash.,
G.P.O., 1945/48-   (Annual. Title varies) 610.72 Un32p

H126. World Health Organization. World Health Statistics Annual.
v.1: Vital statistics and causes of death; v.2: Infectious Diseases:
cases. deaths, vaccinations. v.3: Health personnel and hospital
establishments. Columbia Univ. Pr., 1936/46-   . (Title varies)
614.1 W893s

BIOGRAPHIES

H127. American Men of Medicine. 3d ed., N.Y., Institute for
Research in Biography, 1st ed., 1945-   926.1 W621
   See: Wi-EJ90.

H128. Kelley, E.C. Encyclopedia of Medical Sources. Williams
& Wilkins, 1948. A610 K29e
   See: Wa p. 189.

H129. New York Academy of Medicine. Library. Catalog of
Biographies. G.K. Hall, 1960.
   See: Wi-EJ87.

H130. Thornton, J.L. et al. A Select Bibliography of Medical
Biography. London, The Lib. Assn., 1961. 016.9261 T51s
   See: Wi-EJ83.

DIRECTORIES

H131. American Dental Directory. Chicago, American Dental
Assn., 1st ed., 1947-   614.24 Am35.
   See: Wi-EJ104.

H132. American Medical Directory. Chicago, American Medical
Assn., 1st ed., 1906- . q610 Am3
   See: Med. Lib. Assn. Bull. 56:111-2, 1968. Wi-EJ66

H133. Association of American Medical Colleges. Medical
School Admission Requirements, U.S.A. and Canada. Evanston,
Ill., 1951-   . 610.71 As7a

H134. Directory of Medical Specialists Holding Certification by
American Specialty Boards. Chicago, Advisory Board of Medical
Specialties, 1st ed., 1939- . 610 D629
  See: Wi-EJ91.

H135. Directory of Approved Internships and Residencies.
Chicago, American Medical Assn., 1947- . (Title varies)
610.71 D628
  Published as section in Journal of the American Medical Association 1947-1960.

H136. Health Organizations of the United States, Canada, and
Internationally. 2d ed. Graduate School of Business and Public
Administration, Cornell University, 1965. 610.6 HEA
  See: Wi-1EJ26.

H137. Hospitals: Guide Issue. Chicago, American Hospital Assn.,
1945-   (Title varies) 362.1 H794

H138. Medical Library Association. Directory. Chicago, 1st ed.,
1950. (Publisher varies). 027.9 M4891d
  See: Wi-EJ69.

H139. U.S. Public Health Service. Bureau of State Services.
Directory of State and Territorial Health Authorities. Wash.,
G.P.O., 1913-   (Annual) 614.09 Un3

H140. World Directory of Dental Schools. Geneva, World Health
Organization. 1st ed., 1961- . 617.6071 W893
  See: Wi-EJ106.

H141. World Directory of Medical Schools. Geneva, World Health
Organization, 1st ed., 1953- . 610.71 W89w
  See: Wi-EJ64.

  See also: Uhler, K.L. Foreign Medical Directories. Med. Lib.
Assn. Bull. 48:44-78, 1960.

SERIALS

H142. Andrews, T. World List of Pharmacy Periodicals. Wash.,
American Society of Hospital Pharmacists, 1963.
  See: Med. Lib. Assn. Bull. 51:448, 1963. Wa p. 194.

H143. New York. Medical Library Center of New York. Union
Catalog of Medical Periodicals I. 2d ed., N.Y., 1968.
   See: Med. Lib. Assn. Bull. 56:219-20, 1968.

H144. Pings, V.M. A Plan for Indexing the Periodical Literature of Nursing. N.Y., American Nurses Foundation, 1966.
610.73 P64p

H145. Vital Notes on Medical Periodicals. Chicago, Medical Library
Assn., v.1, 1952-   . A610.5 V83
   See: Wi-EJ22. Wa p. 186.

H146. World Medical Periodicals. 3d ed., N.Y., World Medical
Assn., 1961. (Reprinted by British Medical Journal, 1968)
Supplement, 1968 published by British Medical Journal.
016.6105 W893
   See: Med. Lib. Assn. Bull. 56:538, 1968. Wi-EJ24. Wa p. 186.

   See also:
      Biomedical Serials, 1950-1960. (No. G73)
      Raisig, L. World Biomedical Journals, 1951-60.
      Med. Lib. Assn. Bull. 54:108-25, 1966.
      Lists of periodicals indexed in:
       Biological Abstracts
       Chemical Abstracts
       Excerpta Medica
       Index Medicus

CURRENT INFORMATION; SELECTION AIDS

   Consult such literature as:

      American Hospital Association. Library. The Health Care
         Researcher; Tools of His Trade. Chicago, 1968.

      Brandon, A.N. Selected List of Books and Journals for the
         Small Medical Library. Med. Lib. Assn. Bull.
         57:130-49, 1969.
         Quarterly supplements published in the Bulletin.

      Columbia University Medical Library. Recent Acquisitions:
         a Selected List and Medical Reference Notes. Series 3,
         No. 1, 1965-

Concordia, Sister Mary. Basic Book and Periodical List:
  Nursing School and Small Medical Library. Peru, Ill.
  St. Bede Abbey Pr., 1967.

Council of National Library Associations. Basic List of
  Guides and Information Sources for Professional and
  Patients' Libraries in Hospitals. 1966.

Flandorf, V.S. Medical Books for the Public Library.
  Lib. J. 91:4718-23, 1965.

Medical Books in Print. San Francisco, J.W. Stacey, 1955-

U.S. Veterans Administration. Basic List of Books and
  Journals for Veterans Administration Medical Libraries.
  Wash., 1967. (G-14, M-2, Part XIII, revised)

Catalogs of publications (books, pamphlets, films) of:
  American Dental Association
  American Hospital Association
  American Medical Association
  U.S. National Library of Medicine
  U.S. Public Health Service
  World Health Organization

Current issues of such periodicals as:
  American Medical Association. Journal, v.1, 1883- .
  American Dental Association. Journal, v.1, 1913- .
  American Journal of the Medical Sciences, v.1, 1820-
  Annals of Internal Medicine, v.1, 1922- .
  Journal of Medical Education, v.1, 1926- .
  Lancet, v.1, 1823- .
  Medical Library Association. Bulletin, v.1, 1911- .
  Medical Technology, v.1, 1967- .
  New England Journal of Medicine, v.1, 1812- .
  Today's Health, v.1, 1923- .

SECTION J. AGRICULTURAL SCIENCES

The agricultural sciences is that group of sciences which deals with the practices, arts, sciences, and industries utilized by man to obtain food from the land. The major subdivisions are:

Agricultural economics: concerned with the management and operation of individual farming units as well as the general economic relationship of agriculture to other fields.

Agricultural engineering: application of engineering to agriculture in all of its branches, including buildings, drainage, irrigation, rural electrification, etc.

Agronomy: branch that treats of the principles and practices of crop production and field management.

Animal science: concerned with breeding, feeding, care, and management of animals and the marketing and processing of animals and their products.

Dairy science: deals with the breeding and feeding of dairy animals and the production and utilization of milk and milk products.

Food technology: concerned with the processing of agricultural commodities into finished foods outside the home.

Forestry: deals with the development, protection, and management of forests.

Horticulture: includes flower, fruit, and vegetable cultivation as well as landscape gardening.

Veterinary medicine: study of the prevention and cure of diseases in animals.

Home economics: this organized body of knowledge, usually academically classed as a division of agriculture, treats of food, clothing, shelter, and household management in their physical, economic, and social aspects.

GUIDES TO THE LITERATURE

J1. Blanchard, J.R. & Ostvold, H. Literature of Agricultural Research. Univ. of Calif. Pr., 1958. 016.63 B611
   See: Wi-EK1.

J2. Lauche, R. World Bibliography of Agricultural Bibliographies. München, Bayerische Land., 1957. (Indexes in German and English) 016.63 L36i
  See: Wi-EK5.

J3. Parker, D. et al. Primer for Agricultural Libraries. Oxford, International Association of Agricultural Libraries and Documentalists, 1967.
  See: Coll. & Res. Libs. 29:517, 1968. Lib. J. 93:2830, 1968.

J4. U.S. Department of Agriculture. List of Available Publications of the United States Department of Agriculture. 1929-   .
(Updated by Bimonthly List of Publications and Motion Pictures, 1897-   .) 016.63 Un361
  See: Wa p. 296.

J5. U.S. Department of Agriculture. Motion Pictures of the U.S. Department of Agriculture. 1960-   (Updated by Bimonthly List of Publications and Motion Pictures, 1897-   .) 630 Un3pm

  See also titles listed in the section on BIOLOGICAL SCIENCES

BIBLIOGRAPHIES, INDEXES, ABSTRACTS

  General

    See:
      Biological & Agricultural Index (No. G9)
      Downs, R.B. & Jenkins, F.B. eds. Bibliography. p. 542-57
      (No. A3)

J6. Agricultural Index, Subject Index to a Selected List of Agricultural Periodicals and Bulletins, v.1-49, 1919-64. Continued by Biological & Agricultural Index (No. G10) A630 Ag829
  See: Wi-EK23. Wa p. 294.

J7. International Association of Agricultural Librarians and Documentalists. Quarterly Bulletin. v.1, 1956-   027.905 INT
  See: Wa p. 297 (note).

J8. U.S. National Agricultural Library. Bibliography of Agriculture. Wash., v.1, 1942- . (name of library varies) A630 Un3bi
   See: Report of Task Force ABLE (No. J58). Sp. Libs. 59:712-7, 1968. Wi-EK9.

J9. U.S. National Agricultural Library. Serial Publications Indexed in Bibliography of Agriculture. rev. ed., Wash., 1965. A630 Un3411 no. 75
   See: Wa p. 299 (note).

J10. U.S. National Agricultural Library. Dictionary Catalog of the National Agricultural Library, 1862-1965. v.1, 1968- . (68v. projected) 016.63 N213

J11. National Agricultural Library Catalog. v.1, 1966- . Monthly. 016.63 N213
   Supplements Dictionary Catalog of the National Agricultural Library, 1862-1965 (No. J10).
   See: Wi-1EK2

J12. U.S.D.A. Index to Department Bulletins, nos. 1-1500. G.P.O., 1936. 630 Un3ab
   See: Wi-EK17.

J13. U.S.D.A. Index to Farmers' Bulletins, nos. 1-1750. G.P.O., 1920-41. 630 Un3f
   See: Wi-EK19.

J14. U.S.D.A. Index to Technical Bulletins, nos. 1-750. G.P.O., 1937-41. 2v. 630 Un3t.
   See: Wi-EK18.

J15. U.S.D.A. Index to Publications of the U.S. Department of Agriculture, 1901-1940. G.P.O., 1923-43. 4v. A630.73 Un3li
   See: Wi-EK13.

J16. Velasquez, P. & Nadurille, R. comp. Obras de Consulta Agricolas en Español. Mexico, Biblioteca Agricola Nacional, 1967.

   See also:
      Bulletin Signaletique. 18: Sciences agricules. (No. A18)
      Biological Abstracts. (No. G7)
      Chemical Abstracts. (No. D14)
         See: J. Chem. Doc. 8:98-105, 1968.

149

J17. Abstracts of Literature on Milk and Milk Products. 1936- .
(Published monthly in Journal of Dairy Science) 637.05 JO

J18. Journal of the Science of Food and Agriculture. v.1, 1950 .
(Abstract section covers chemistry of agriculture, food and sani-
tation) 630.1605 JO
  See: Wi-ED16. Wa p. 362.

J19. Pesticides Documentation Bulletin. National Agricultural
Library. v.1, 1965. (biweekly) 016.632 Un28p
  See: Lib. J. 90:3572, 1968.

J20. World Agricultural Economics and Rural Sociology Abstracts,
v.1, 1959- . 305 WORLD
  See: Wa p. 297.

    Commonwealth Agricultural Bureaux publications:

J21. Animal Breeding Abstracts. v.1, 1933-   636.05 IMP
  See: Wa p. 315.

J22. Bibliography of Soil Science, Fertilizers and General Agronomy.
v.1, 1931- . A631 Im7b
  See: Wi-EK2; 1EK1. Wa p. 295.

J23. Dairy Science Abstracts. v.1, 1939-   637.05 DAIS
  See: Wi-EK26. Wa p. 317.

J24. Field Crop Abstracts. v.1, 1948-   633.05 FI
  See: Wa p. 309.

J25. Herbage Abstracts. v.1, 1931-   634.605 HE
  See: Wa p. 309.

J26. Horticultural Abstracts. v.1, 1931-   634.05 IMP
  See: Wa p.312.

J27. Nutrition Abstracts and Reviews. v.1, 1931/32-
612.3905 NUA
  See: Wa p. 190.

J28. Plant Breeding Abstracts. v.1, 1930- . 634.05 PL
  See: Wa p. 307.

J29.   Review of Applied Mycology. v. 1, 1922-   589. 205 RE
  See: Wa p. 160.

J30.   Soils and Fertilizers. v. 1, 1938-   . 631. 05 IM
  See: Wa p. 295 (note).

J31.   Weed Abstracts. v. 1, 1952-   632. 705 WEE
  See Wa p. 308.

HISTORIES

J32.   Brooklyn Botanic Garden.   Origins of American Horticulture:
a Handbook. Brooklyn, 1968.

J33.   Carrier, L.   The Beginnings of Agriculture in America.
McGraw-Hill, 1923.   Reprint, Johnson Reprint, 1968.   630. 973 C23b

J34.   Edwards, E. E.   A Bibliography of the History of Agriculture
in the United States. Wash., G. P. O., 1930.   Reprint, Gale, 1967.
630 Un3mi no. 84.
  See: RQ 7:182, 1968.

J35.   Gras, N. S. B.   History of Agriculture in Europe and America.
2d ed., Crofts, 1940.   Reprinted, 1946.   630. 9 G76h2

J36.   Gray, L. C.   History of Agriculture in the Southern United
States to 1860 1933. 2v.   Carnegie Institute of Wash.,   Pub. 430.
Reprinted 1941.   630. 973 G79h

J37.   Bidwell, P. W.   History of Agriculture in the Northern United
States, 1620-1860.   1925.   (Carnegie Institute of Wash., Pub. 358)
Reprinted 1941.   630. 973 B47h

J38.   Moore, E. G.   The Agricultural Research Service.   Praeger,
1967.   353. 81 M78a
  See: Science 159:1221-2, 1968.

J39.   Rasmussen, W. D. ed.   Readings in the History of American
Agriculture.   Univ. of Ill. Pr., 1960.   630. 973 R184r

J40.   Taylor, H. C. & Taylor, A. D.   Story of Agricultural Economic
in the United States, 1840-1932.   Iowa State College Pr., 1952.
338. 1 K21s

J41. U.S.D.A. A Guide to Understanding the United States Department of Agriculture. G.P.O., 1965.

J42. U.S. Department of Agriculture. Economic Research Service. A Preliminary List of References for the History of Agricultural Science and Technology in the U.S. Davis, Univ. of California, 1966. 016.630973 P97p

DICTIONARIES

J43. Dictionary of Agricultural and Allied Terminology. Mich. State Univ. Pr., 1962. 630.3 D56.

J44. Haensch, G. & Haberkamp, G. comp. Dictionary of Agriculture: German/English/French/Spanish. 2d ed., Elsevier, 1963. 630.3 H11w.
  See: Lib. J. 88:2677, 1963. Wa p. 298.

J45. Usovsky, B.N. et al. Comprehensive Russian-English Agricultural Dictionary. 1st English ed., Pergamon, 1967. (Based on 1960 Russian edition) 630.3 Us6r
  See: Wa p. 298.

  See also: U.S. National Agricultural Library. Agricultural/Biological Vocabulary. (No. G30)

HANDBOOKS, DESCRIPTIVE WORKS

J46. Association of Official Agricultural Chemists. Official Methods of Analysis. Wash., 1st ed., 1919- . (Title varies) 630.16 As7o
  Updated by its Changes in Official Methods of Analysis made at the annual meeting and reprinted from its Journal, 1956-

J47. Doane Agricultural Service, Inc. Farm Building Cost Handbook. St. Louis, 1963. (looseleaf) 630.2 D65fa

J48. Ensminger, M.E. Beef Cattle Science. 4th ed., Interstate, 1968. 636 En7b

J49. Farm Chemical Hand Book. v.1, 1908-   (title varies) 668.6 Am3

D50. Frandsen, J.H. et al. Dairy Handbook and Dictionary.
Amherst, Mass., 1958. 637 F85d
  See: Science 129:1422, 1959.

D51. Mallis, A. Handbook of Pest Control. MacNair-Dorland,
1964. 648.7 M29h

J52. Midwest Farm Handbook. Iowa State Univ. Pr., 1st ed.,
1949-   . 630 M584.

J53. Morrison, F.B. Feeds and Feeding, a Handbook for the
Student and Stockman. 22d ed., Ithaca, Morrison, 1956. 635 H39f
  Also available in an abridged edition.

J54. Pesticide Handbook - Entoma 1968. 20th ed., College Science
Pubs., 1968. 632.4 P439

J55. U.S. Department of Agriculture. Yearbook of Agriculture.
1894-   . (annual) 630 Un3al
  Since 1936 statistics are published as Agricultural Statistics
(No. J63) and the Yearbook consists of comprehensive treatment
of a special subject:

1936	Better Plants and Animals I
1937	Better Plants and Animals II
1938	Soils and Men
1939	Food and Life
1940	Farmers in a Changing World
1941	Climate and Man
1942	Keeping Livestock Healthy
1943-7	Science in Agriculture
1948	Grass
1949	Trees
1950-1	Crops in Peace and War
1952	Insects
1953	Plant Diseases
1954	Marketing
1955	Water
1956	Animal Diseases
1957	Soil
1958	Land
1959	Food
1960	Power to Produce
1961	Seeds
1962	After a Hundred Years
1963	A Place to Live
1964	The Farmer's World

```
1965    Consumers All
1966    Protecting Our Food
1967    Outdoors U.S.A.
1968    Science for Better Living.
```

See: Wi-EK37. Wa p. 300.

J55a. U.S. Department of Agriculture. Summary of Registered Agricultural Pesticide Chemical Uses. 3d ed. Wash., G.P.O., 1968- . 632.4 Un312s

J56. Westcott, C. Plant Disease Handbook. 2d ed., Van Nostrand, 1960. 632.6 W52pl

INFORMATION ACTIVITIES

See BIOLOGICAL SCIENCES and MEDICAL SCIENCES

J57. Finney, D.J. An Introduction to Statistical Science in Agriculture. 2d ed., Copenhagen, Munksgaard, 1962. 311 F49i

J58. U.S. National Agricultural Library. Report of Task Force ABLE: Agricultural Biological Literature Exploitation; a System Study of the National Agricultural Library and its Users. Wash., 1965. 010.78 Un312a

See also: U.S. National Agricultural Library. Agricultural/ Biological Vocabulary. (No. G32)

AGRICULTURAL STATISTICS

J59. Illinois Agricultural Statistics. Assessors' Annual Farm Census. Springfield, Ill., Cooperative Crop Reporting Service, 1937- . (title varies) 338.1 Il661if

J60. Production Yearbook. Rome, F.A.O., v.12, 1958- (Continued and took volume numbering of Yearbook of Food and Agricultural Statistics. Part 1: Production, 1947-57) 338.1 P945

J61. Trade Yearbook. Rome, F.A.O., v.12, 1958- . (Continued and took volume numbering of Yearbook of Food and Agricultural Statistics. Part 2: Trade) 382 T675
See: Wi-CH164.

J62. U.S. Bureau of the Census. 1964 United States Census of Agriculture. Wash., 1967-8. 3v. q630.973 Un321c

J63. U.S. Department of Agriculture. Agricultural Statistics. Wash., 1936- . (Previously published in U.S.D.A. Yearbook of Agriculture, 1894-1935 (No. J52)) 630 Un3agr.
  See Wi-EK38.

J64. U.S. Department of Agriculture. Economic Research Service. Foreign Agriculture including Foreign Crops and Markets. Wash., v.1, 1963. (monthly) 338.105 FOA

J65. U.S. Department of Agriculture. Foreign Agricultural Service. The World Agricultural Situation. G.P.O., 1956- . (Superseded World Food Situation) 338.1 Un3283w

J66. U.S. Department of Agriculture. Foreign Agricultural Service. World Agricultural Production and Trade, Statistical Report, 1963- . (Superseded World Summaries, Crop and Livestock, 1957-62) 338.105 UNIS3

J67. U.S. Department of Agriculture. Handbook of Agricultural Charts. Wash., G.P.O., 1963- . (Title varies) 338.1 Un3h

  See also:
    U.S. Department of Agriculture. Economic Research Service. Situation Reports. A series of reports published 3 to 5 times a year. Includes Cotton Situation; Demand & Price Situation; Farm Income Situation; Fats and Oils Situation; Feed Situation; Fruit Situation; Marketing & Transportation Situation; National Food Situation; Livestock & Meat Situation; Poultry Situation; Tobacco Situation; Vegetable Situation; Wheat Situation; Wool Situation.

DIRECTORIES

J68. Directory of Agricultural and Home Economic Leaders of the U.S. and Canada. Cambridge, Mass., W.G. Grant, 1919- . (Annual) 630.31 D628

J69. Stephenson, J.W. The Gardener's Directory. Hanover House, 1960. 634 St4g
  See: Sub. Bks. Bull. 57:202, 1960. Wa p. 314.

J70. U.S. Department of Agriculture. Directory of Organizations and Field Activities of the Department of Agriculture v.1, 1925-30.31 Un3d

J71. U.S. Department of Agriculture. Workers in Subjects Pertaining to Agriculture in the Land-Grant Colleges and Experiment Stations Wash., G.P.O., 1924/25-   . 630.7 Un311i

GARDENING

J72. Abraham, G. The Green Thumb of Indoor Gardening; a Complete Guide. Prentice-Hall, 1967. 634.35 Ab8g
  See: Booklist 64:364, 1967.

J73. Dodge, B.O. et al. Diseases and Pests of Ornamental Plants. 3d ed., Ronald, 1960. 632.3 D66d
  See: Sub. Bks. Bull. 57:118, 1960.

J74. Elbert, G. & Hyams, E.S. House Plants. Funk & Wagnalls, 1968.
  See: Booklist 65:468, 1969.

J75. Foster, H.L. Rock Gardening. Houghton, 1968.
  See: Lib. J. 93:3149, 1968.

J76. New Illustrated Encyclopedia of Gardening. Greystone, 1960. 6v. 634.03 N42
  See: Sub. Bks. Bull. 56:228-30, 1960.

J77. Snyder, R. ed. The Complete Book for Gardeners. Van Nostrand, 1964. 635 Sn92c
  See: Lib. J. 89:1975, 1964.

J78. Taylor's Encyclopedia of Gardening, Horticulture, and Landscape Design. 4th ed., Houghton, 1961. 634.03 T21e
  See: Science 133:1240, 1961. Sub. Bks. Bull. 57:685-6, 1961.

J79. Taylor, N. The Guide to Garden Shrubs and Trees. Houghton, 1965. 634.85 T22g
  See: Booklist 63:646-7, 1966.

FORESTRY

Bibliographies, Indexes, Abstracts

J80. Forestry Abstracts. Farnham Royal, England, Commonwealth Agricultural Bureaux, v.1, 1939-   . 634.905 FORA
  See: Wi-EK44. Wa p. 309-10, 389.

J81. Yale University. School of Forestry Library. Dictionary
Catalogue of the Henry S. Graves Memorial Library. G. K. Hall,
1962. 12v.

Dictionaries & Encyclopedias

J82. Hough, R. B. Hough's Ecncylopedia of American Woods.
N. Y., Speller, 1957-   (15v. of text and specimens) 581.6 H81a

J83. Panshin, A. J. et al. Textbook of Wood Technology. 2d ed.,
McGraw-Hill, 1964- . 634.9 B813i
  See: Wa p. 389.

J84. Society of American Foresters. Forestry Terminology; a
Glossary of Technical Terms used in Forestry. 3d ed., Wash.,
1958.
  See: Wi-EK48.

J85. Elsevier's Wood Dictionary in Seven Languages: American,
French, Spanish, Italian, Swedish, Dutch, and German. Elsevier,
1964- . 634.903 E176
  See: Wa p. 390.

J86. Weck, J. comp. Dictionary of Forestry in Five Languages:
German, English, French, Spanish, Russian; with Appendices of
Tree Species and Animals and Plants Causing Forest Pests and
Disease. Elsevier, 1966.
  See: New Tech. Bks. 52:176, 1967. Lib. J. 92:104, 1967.
Wi-1EK4.

Handbooks, Atlases, Yearbooks

J87. Davis, K. P. Forest Management: Regulation and Valuation.
2d ed., McGraw-Hill, 1966. 634.9 D29f
  See: J. of Forestry 65:658, 1967.

J88. Forbes, R. D. ed. Forestry Handbook. Ronald, 1955.
634.9 F74f
  See: New Tech. Bks. 40:97, 1955. Wi-EK48.

J89. Little, E. L. Checklist of Native and Naturalized Trees of
the U. S. Wash., 1953. (U.S. D. A. Agricultural Handbook No. 41)
582 L72c

J90. Reinbek, Germany. Bundesanstalt für Forst- und Holzwirt-
schaft. Weltforstatlas. (World Atlas of Forestry) Berlin, Haller,
1951- . f634.9 R274w
  See: Wa p. 311.

J91. U.S. Department of Agriculture. Forest Service. Silvics of
Forest Trees of the U.S. Wash., 1965. (Agriculture Handbook No.
271)

J92. Yearbook of Forest Products Statistics. Rome, F.A.O., v. 1,
1947- . 338.1 Y33
  See: Wa p. 312.

VETERINARY MEDICINE

  Indexes, Abstracts, Reviews

J93. Advances in Veterinary Sciences. Academic, v. 1, 1953- .
619 Ad95
  See: Wa p. 202.

J94. Helminthological Abstracts. Farnham Royal, England, Commonwealth Agricultural Bureaux. v. 1, 1932- . 595.105 HE
  See: Wa p. 166.

J95. Index Veterinarius. Weybridge, Surrey, England, Imperial
Bureau of Animal Health. v. 1, 1933- . 619.05 IND
  See: Wi-EJ145. Wa p. 201 (note).

J96. Veterinary Annual. Bristol, John Wright, v. 1, 1959- .
619 V638
  See: Wa p. 202.

  See also: Index-Catalogue of Medical and Veterinary Zoology
    (No. G187)

  Handbooks, Manuals, Yearbooks

J97. American Kennel Club. The Complete Dog Book. Rev. ed.,
Garden City, 1961. 636.7 Am3p

J98. Animal Health Yearbook. Rome, F.A.O., 1957- .
q619 An543
  See: Wa p. 202.

158

J99. Barnes, C.D. & Eltherington. Drug Dosage in Laboratory Animals; a Handbook. Univ. of Calif. Pr., 1964. 615.14 B26d
  See: Science 148:65-6, 1965.

J100. Guthrie, E.L. & Miller, R.C. Home Book of Animal Care. Harper, 1966. 636 G98h
  See: A.L.A. Booklist 63:980, 1966. Lib. J. 91:1435, 1966.

J101. The Merck Veterinary Manual. Rahway, Merck. 1st ed., 1955-   . 636.0896 M537
  See: Wi-EJ146. Wa p. 202.

J102. Miller, W.C. & West, G.P. Encyclopedia of Animal Care. 8th ed., Williams & Wilkins, 1967. (Formerly Black's Veterinary Dictionary) 619 M61b
  See: Wi-EJ147. Wa p. 201 (note).

J103. Veterinarians' Blue Book. R. H. Donnelley, 1st ed., 1953- (Formerly Veterinary Drug Encyclopedia and Therapeutic Index, 1953-1966) 619.03 V641
  See: Wi-EJ148.

J104. Year Book of Veterinary Medicine. Year Book Medical Pub., v.1, 1963-65. 636.089 Y32

  Directories

J105. American Veterinary Medical Association. Directory. Chicago, 1st ed., 1924-   . (title varies) 619 Am35d

J106. World Directory of Veterinary Schools. Rome, F.A.O., 1st ed., 1963-   636.089071 W893
  See: Wa p. 202.

  See also:
    Medical Library Association. Library Statistics of Veterinary Schools in the United States and Canada. Med. Lib. Assn. Bull. 55:201-6, 1967.

HOME ECONOMICS

  Bibliographies, Indexes, Abstracts, Reviews

  See also sections BIOLOGICAL SCIENCES, CHEMISTRY, and
    MEDICAL SCIENCES.

J107. Advances in Food Research. Academic, v. 1, 1948-
641.1. Ad95

J108. Baker, E.A. Bibliography of Food: a Select International
Bibliography of Nutrition, Food and Beverage Technology and Dis-
tribution, 1936-56. Academic, 1958. 016.641 B17b
  See: Wa p. 362.

J109. Commonwealth Agricultural Bureaux. Evaluation of the World
Food Literature; Results of an International Survey. Farnham
Royal, England, 1967.
  See: Food Technology 27:1612, 1967.

J110. Gunderson, F.L. et al. Food Standards and Definitions in
the United States: a Guidebook. Academic, 1963. 614.3 G95f

J111. Lincoln, W. American Cookery Books, 1742-1860. N.Y.,
Corner Book Shop, 1954. A641.5 L64a
  See: Wi-EK53.

Foods

J112. The Almanac of the Canning, Freezing, Preserving Industries. Westminster, Md., E.E. Judge, v.1, 1916. (Annual; title
varies) 664.8 C16a

J113. Bender, A.E. Dictionary of Nutrition and Food Technology.
3d ed., Wash., Butterworths, 1968. 641.03 B43d
  See: Wi-1EJ23.

J114. Bender, A.E. Dietetic Foods. Chemical Pub., 1968.
613.2 B432d

J115. Borgstrom, G. Principles of Food Science. Macmillan,
1968. 2v. v.1: Food Technology; v.2: Food Microbiology and
Biochemistry. 664 B64p
  See: Lib. J. 94:529, 1969; Choice 6:243, 1969.

J116. Cox, H.E. & Pearson, D. The Chemical Analysis of
Foods. Chemical Pub., 1962. 543.1 C83c

J117. Field, H.E. Foods in Health and Disease: A Practical
Guide. Macmillan, 1964. 613.2 F45f
  See: Lib. J. 89:1973, 1964.

J118. Food Chemical News. Guide to the Current Status of Food Additives and Color Additives. Wash., L. Rothchild Jr., 1967. (looseleaf) 614.31 F73g

J119. Handbook of Food Additives. Chemical Rubber Co., 1968. 664 C859

J120. IFI-Plenum Data Corporation. Directory, 1965-8. N.Y., 1968. (Looseleaf; annual revision service)

J121. IFI-Plenum Data Corporation. Food and Color Additives. Index. (Cumulative through 1968 in looseleaf format and revision service) N.Y., 1968.

J122. Laboratory Handbook of Methods of Food Analysis. Chemical Rubber Co., 1968.

J123. Lichine, A. et al. Alexis Lichine's Encyclopedia of Wines and Spirits. Knopf, 1967.
  See: Lib. J. 92:4156, 1967.

J124. National Academy of Science-National Research Council. Recommended Dietary Allowances. 6th ed., Wash., 1964. 613.2 N213r

J125. Rombauer, I.S. & Becker, M.R. The Joy of Cooking. Bobbs-Merrill, 1962. 641.5 R66j

J126. U.S. Department of Agriculture. Composition of Foods: Raw, Processed, Prepared. Wash., 1963. (Agricultural Handbook No. 8) 641.1 W34c

J127. U.S. Panel on the World Food Supply. The World Food Problem; a Report. Wash., G.P.O., 1967. 6v. 338.19 Un334w

  See also:
    Food, the Yearbook of Agriculture, 1959 (No. J52)
      See: Science 132:952, 1960.
    Lowry, A.D. & Cocroft, R. Literature Needs of Food Scientists.
      J. Chem. Doc. 8:228-30, 1968.
    Marcus, I. Agriculture and Food Technology in the Patent
      Office. J. Chem. Doc. 8:225-7, 1968.
    Piskur, M.M. The Literature of Food Science and Technology.
      J. Chem. Doc. 8:93-5, 1968.
    Skolnik, H. History, Evolution, and Status of Agriculture and
      Food Science and Technology. J. Chem. Doc. 8:95-8, 1968.

Textiles

J128. American Home Economics Association. Textile Handbook.
Wash., 1st ed. 1960-   .   (3d ed., 1966) 677.02 Am3t

J129. Encyclopedia of Textiles. 2d ed., Prentice-Hall, 1966.
q677 Am32a
  See: Sp. Libs. 58:127, 1967.

J130. Fairchild's Dictionary of Textiles. 2d ed., N.Y.,
Fairchild Pubs., 1967. 677.03 F16
  See: Lib. J. 93:742, 1968.

J131. Hamby, D.S. ed. The American Cotton Handbook. 3d ed.,
Interscience, 1965. 2v. (Earlier editions by G.R. Merrill)
677 M55a
  See: New Tech. Bks. 52:70, 1967.

J132. Klapper, M. Fabric Almanac. N.Y., Fairchild Pubs.
1st ed., 1966-   . 677.02 K66f

J133. Linton, G.E. Modern Textile Dictionary. Duell, 1954.
677.03 L65m
  See: Sub. Bks. Bull. 26:56, 1955. Wa p. 396 (note).

J134. Mark, H.F. et al. Man-made Fibers; Science and Technology. Interscience, v.1, 1967- . 677.4 M34m
  See: Am. Scientist 56:299A, 1968.

J135. Natural and Synthetic Fibers Abstract Service. Interscience,
1954- . (monthly, looseleaf)
  Compiled annually as Natural and Synthetic Fibers Yearbook.

J136. Polanyi, M. Technical Trade Dictionary of Textile Terms:
German-American/English; American/English-German. 2d ed.,
Pergamon, 1967.
  See: Choice 4:1228, 1968; RQ 7:49, 1967.

J137. Review of Textile Progress. Plenum, v.1, 1949-   .
(Annual) 677 R325
  See: Wa p. 397.

J138. World Textile Abstracts. Didsbury, England, Shirley
Institute. v.1, 1969- . (Semi-monthly) 677.05 WOR
  See: Chem. & Eng. News 47:52, (Mar. 17) 1969.

Handbooks for the Layman

J139. Daniels, G. E. Home Guide to Plumbing, Heating, and Air Conditioning. Popular Science Pub., 1967.
  See: Booklist 64:970-1, 1967. Lib. J. 92:2170, 1967.

J140. Good Housekeeping. Good Housekeeping's Guide for the Young Homemakers; an Up-to-the-minute Handbook of Successful Home Management. Harper, 1966. 640 G58go
  See: Booklist 64:466, 1967.

J141. Harmon, A. J. The Guide to Home Remodeling. Holt, 1967. 643.7 H22g
  See: Booklist 64:971, 1967.

J142. Hertzberg, R. E. Repairing Small Electrical Appliances. Arco, 1968.
  See: Booklist 65:621, 1969.

J143. Moore, A. C. How to Clean Everything; an Encyclopedia of What to Use and How to Use it. 3d ed., Simon & Schuster, 1968.
  See: Lib. J. 94:179, 1969.

J144. O'Neill, B. P. & O'Neill, R. W. The Unhandy Man's Guide to Home Repairs; a Complete Guide to Home Maintenance, Improvement, and Remodeling, for Men and Women. Macmillan, 1966.
  See: Booklist 64:704, 1967.

J145. The Practical Handyman's Encyclopedia: The Complete Illustrated Do It Yourself Library for Home and Outdoors. Greystone, 1965. 18v.
  See: Sub. Bks. Bull. 64:1241-4, 1968.

J146. Vanderbilt, A. Amy Vanderbilt's New Complete Book of Etiquette; a Guide to Gracious Living. 6th ed., Doubleday, 1963.

J147. Whitman, R. B. First Aid for the Ailing House. 5th ed., McGraw-Hill, 1958. 643.7 W59f

  See also: Consumers All, Yearbook of Agriculture, 1965 (No. J52)

Directories

J148. U. S. Public Health Service. Directory of Homemaker Services: Homemaker Agencies in the United States. Wash., G. P. O., 1958- . 362.82 Un32d

SERIALS

See: U.S. National Agriculture Library. Serial Publications Indexed in Bibliography of Agriculture (No. J9)

J149. International Association of Agricultural Librarians and Documentalists. Current Agricultural Serials; a World List of Serials in Agriculture and Related Subjects (excluding Forestry and Fisheries) Current in 1964. Oxford, Alden, 1965-67. 2v.
  Supplemented by "New Agricultural Serials" in Quarterly Bulletin of the Association. 016.6305 In8c
  See: Coll. & Res. Libs. 27:310. 1966. Wi-1EK3. Wa p. 299.

CURRENT INFORMATION: SELECTION AIDS

Consult current issues of such periodicals as:
    Agricultural Education Magazine, v.1, 1929- .
    American Dairy Review, v.1, 1939- .
    American Dietetics Association. Journal, v.1, 1925- .
    American Forests, v.1, 1895- .
    American Journal of Agricultural Economics, v.1, 1919- .
    American Veterinary Medicine Association. Journal, v.1,
    . 1869- .
    Food Technology, v.1, 1947- .
    The Garden Journal, v.1, 1951- .
    Journal of Forestry, v.1, 1902- .
    Nutrition Reviews, v.1, 1942- .
    Pesticides Monitoring Journal. v.1, 1967- .
    Prairie Farmer, v.1, 1841- .
    Soil Science, v.1, 1916- .
    What's New in Home Economics, v.1, 1937- .

Lists of publications of:
    U.S. Department of Agriculture
    U.S. Public Health Service
    Food and Agriculture Organization (FAO)
    World Health Organization (WHO)

SECTION K. ENGINEERING SCIENCES

The engineering sciences is that group of sciences by which the properties of matter and the sources of power are made useful to man in
structures, machines, and manufactured products.   The major fields
of engineering are:
  Aeronautical and astronautical engineering: concerned with the
      study, design, development, manufacture, and operation of
      airplanes, missiles, and space vehicles, and with navigation in
      space.
  Agricultural engineering: deals with the application of the funda-
      mentals of engineering to the peculiar conditions and require-
      ments of agriculture.
  Ceramic engineering; concerned with the scientific and engineering
      principles involved in the investigation, production, and utiliza-
      tion of the solid state articles of commerce and science called
      ceramic materials.
  Chemical engineering: involves the development and application of
      manufacturing processes by which raw materials are changed to
      products for industrial and commercial use.
  Civil engineering: the division that deals with planning, designing
      construction, and maintenance of a large variety of structures
      and facilities for public, commercial and industrial use.
  Electrical and electronic engineering: deals with applications of
      electrical energy to the benefits of mankind.   Includes such
      areas as ultrasonics, power generation, transmission and dis-
      tribution, vacuum and gas tubes, antennas and fields, microwave
      generation and propogation, network analysis and synthesis,
      computers, transistors, solid state devices, physical electronics,
      etc.
  Mechanical engineering: concerned with the theory of conversion
      and transmission of energy and the practical utilization of power
      processes as well as the technological and economic aspects in
      the development, design and utilization of machines and systems.
  Industrial engineering: study of the design, improvement, and
      installation of integrated systems of men, materials, and equip-
      ment.
  Mining and metallurgical engineering: concerned with all of the
      steps in the production of metals, from location of mineral
      deposits, through extraction from natural ores, to marketing
      to meet man's needs.
  Nuclear engineering: deals with the control and utilization of
      energy and radiation from nuclear sources.

Petroleum engineering: interested in the engineering aspects of
the rock and rock-fluid systems of the earth's crust, particularly
the principles of oil production.
Sanitary engineering: application of engineering and biological
principles to man's environment for the protection of public
health.

GUIDES TO THE LITERATURE

General

K1. Dalton, B.H. Sources of Engineering Information. Univ. of
Calif. Pr., 1948. A620 D17s
 See: **Wi-EI1.** Wa p. 203.

K2. Sternberg, V.A. How to Locate Technical Information. Prentice-
Hall, 1964. 016.6 St8h

Special

K3. Blaisdell, R.F. et al. Sources of Information in Transportation.
Northwestern Univ., 1964. 016.385 Ad4s

K4. Burkett, J. & Plumb, P. How to Find Out in Electrical
Engineering. Pergamon, 1967. 621.3 B91h

K5. Fry, B.M. & Mohrhardt, F.E. eds. A Guide to Information
Sources in Science and Technology. Interscience, 1963. (Guides
to Information Sources in Science and Technology. v.1) 016.6294
F9g
 See: Wi-EI9. Wa p. 289.

K6. How-to-do-it Books; a Selected Guide. 3d ed., Bowker,
1963. (First two editions compiled by R.E. Kingery) 016.6
K59h
 See: Wi-CB156. Wa p. 417.

K7. Smith, D.L. How to Find Out in Architecture and Building.
Pergamon, 1967. 016.72 Sm5h
 See: New Tech. Bks. 52:278, 1967.

K8. Smith, F.S. Know-how Books; an Annotated Bibliography of
Do It Yourself Books. Bowker, 1957. 028 S645k
 See: Wi-CB158. Wa p. 417 (note).

K9. Special Libraries Association. Guide to Metallurgical Information. 2d ed., N.Y., 1965. (S.L.A. Bibliography, No. 3)
010 Sp3s no. 3

K10. Yescombe, E. R. Sources of Information on the Rubber,
Plastics and Allied Industries. Pergamon, 1968. (International
Series of Monographs in Library and Information Science, v.7)
016.6684 Y48s

   See also: Fetros, J.G. How-To-Do-It Books. <u>Lib. J.</u>
94:1595-7, 1969

INDEXES & ABSTRACTS

   See also GENERAL SCIENCE, MATHEMATICS, PHYSICS,
CHEMISTRY and EARTH SCIENCES.

   ### General

K11. Applied Science & Technology Index. H. W. Wilson, v.46,
1958- . (Supersedes in part and continues the numbering of
<u>Industrial Arts Index</u>, v.1-45, 1913-57) 605 IND2
   See: Wi-EA63. Wa p. 4.

K12. British Technology Index. London, Library Assn., v.1,
1962- . 605 BRI
   See: <u>Lib. J.</u> 87:2482-4, 1962. Wi-EA64.

K13. Engineering Index. N. Y., Engineering Index, Inc., v.1,
1884- .
   Coverage: v.1, 1884-1891; v.2, 1892-1895; v.3, 1896-1900;
v.4, 1901-1905. Annual, 1906- . Monthly cumulating into
annual, 1962- . Monthly available on magnetic tape (COMPENDEX),
1969- . Daily and weekly card service, 1928- . On an experimental basis <u>Electrical and Electronic Engineering Section</u>
and <u>Plastics Section</u> available, 1965-68. Magnetic tapes of these
sections available, 1969- . A620 En37
   See: <u>Lib. J.</u> 94:706, 1969. Wa p. 203-4.

K14. Engineering Societies Library. Classed Subject Catalog.
G.K. Hall, 1964. 13v. Supplements, 1964-
   See: Wa p. 204.

K15. Engineers' Council for Professional Development. Selected
Bibliography of Engineering Subjects. N. Y., 1937- . A600 En3s
   These bibliographies of books on engineering subjects include:
   I. Mathematics, Mechanics & Physics, 1962.

II. Aeronautical Engineering, 1950
III. Civil Engineering, 1962
IV. Ceramic Engineering, 1965
V. Metallurgical, Mining and Geological Engineering, 1962
VI. Mechanical Engineering, 1955
VII. Electrical Engineering, 1958
VIII. Chemical Engineering, 1962
IX. Industrial Engineering, 1962

See also:
Bulletin Signalétique. 9, Sciences de l'Ingenieur (No. A18)
Referativnyl Zhurnal. (No. A19)
Downs, R. B. & Jenkins, F. B., eds. Bibliography. p. 530-41.
(No. A3)

Special

K16. Abstracts of Photographic Science & Engineering Literature.
(APSE). N. Y., Society of Photographic Scientists and Engineering
Terms. v.1, 1962- . (Monthly; six-year cumulative index)
770.5 AB

K17. Air University Library Index to Military Periodicals. Maxwell
Air Force Base, Ala., Air University Library, v.1, 1949-
(Title and publisher vary) 629.1305 AIU
See: Wi-EI22.

K18. APCA Abstracts. Pittsburgh, Air Pollution Control Assn.,
v.1, 1955- . 614.7105 AI
See: Wa p. 271.

K19. Applied Mechanics Reviews. Easton, Pa., American Society
of Mechanical Engineers, v.1, 1948- . 620.5 APM
Indexed by WADEX, a separate computer-produced author and
title index. v.15, 1962- .
See: Wa p. 210.

K20. Ashburn, E.V., ed. Laser Literature; a Permuted Bibliography 1958-1966. Western Periodicals Co., 1967. 2v 016.621329
As321

K21. Battelle Technical Review. Columbus, Ohio, Battelle Memorial
Inst., v.1, 1952- . 605 BAT
See: Wa p. 3.

K22. Building Science Abstracts. London, H.M. Stationery Office, v.1, 1928- . 690.5 BUIS
  See: Wa p. 418.

K23. Ceramic Abstracts. Easton, Pa., AmericanCeramic Society. v.1, 1922- . Published as abstract section in the society's Journal. 666.06 AMBA
  See: Wa p. 372.

K24. Codlin, E.M. Cryogenics and Refrigeration; a Bibliographical Guide. Plenum, 1968. 016.62159 C64c

K25. Computer & Control Abstracts (CCE). N.Y., Institute of Electrical and Electronic Engineers. London, Institution of Electrical Engineers.
  Supplemented by Current Papers in Control (CPC)
  See: Wi-1EL25.

K26. Current Literature in Traffic and Transportation. Evanston, Ill., Northwestern University, v.1, 1960- . 385.05 CU

K27. Electrical & Electronics Abstracts (EEA). N.Y., Institute of Electrical & Electronic Engineers; London, Institution of Electrical Engineers. v.6, 1903- . (Continues the numbering of Science Abstracts Section B, v.1-5, 1898-1902) Title varies. 530.5 SA
  Supplemented by Current Papers for the Professional Electrical and Electronics Engineer, v.1, 1964- .
  See: Wi-EI95. Wa p. 225.

K28. Electronics and Communications Abstracts. Brentwood, England, Multiscience Pub., v.1, 1961- . 621.3805 EL
  See: Wa p. 232.

K29. Engineering Plastics Monthly. N.Y., Society of Plastics Engineers. v.5, 1969- . (Supersedes and continues the volume numbering of Engineering Index: Plastics Section, v.1-4, 1965-68) 668.405 PLM

K30. Fuel Abstracts and Current Titles. London, Institute of Fuel, v.1, 1960- . 662.605 FUA
  See: Wa p. 358.

K31. Gas Abstracts. Chicago, Institute of Gas Technology, v.1, 1945- . 665.05 GAA
  See: Wa p. 360.

K32. Highway Research Abstracts. Wash., 1931-   Cumulative index, 1931--61. 1v. (Volume numbering irregular) 625.705 NA
  See: Wi-E184.

K33. International Aerospace Abstracts. Institute of the Aerospace Sciences. v.1, 1961-   . 629.1305 INTE
  See: Wi-EI19. Wa p. 282.

K34. Laser Abstracts. Evanston, Ill., Lowry-Cocroft, 1960- (A weekly edge-notched card service. The abstracts relating to medicine, dentistry and biology are published also by themselves as a separate service)

K35. Laser and Maser International. Vancouver, Canada, Pacific Impex Co., 1964-   . (Looseleaf monthly abstracting service)
  See: Physics Today 17:108, (Jan.) 1964.

K36. The Laser Literature: An Annotated Guide. Plenum, 1968.
    Covers literature 1963-6, continuing Laser Abstracts (Plenum, 1964) which covered literature through mid-1963. 016.621381 L33
  See: Wa p. 231 (note).

K37. Metals Abstracts. Metals Park, Ohio, American Society for Metals. v.1, 1968. 669.05 MEAB
  Formed by a merger of Review of Metal Literature, 1944-67 and Metallurgical Abstracts (1909-67)
  See: Sci. Info. Notes 10:24, (Apr.-May) 1968.

K38. Moore, C.K. & Spencer, K.J. Electronics: a Bibliographical Guide. London, MacDonald; N.Y., Plenum. 1961-   . 016.537 M78e
  See: Wa p. 231.

K39. Public Health Engineering Abstracts. Wash., G.P.O., v.1, 1921-   . 628.05 UNI
  See: Wa p. 270.

K40. Selected RAND Abstracts. Santa Monica, Calif., Rand Corporation, v.1, 1963-   . (Quarterly cumulated, December number constitutes the annual cumulation) 016.6072 R15s
  RAND publications before 1963 are indexed by Index of Selected Publications of the RAND Corporation, 1946-62; Supplement, 1963.

K41. Solar Energy: The Journal of Solar Energy Science and
Engineering. Section: Solar Abstracts. Phoenix, Assn. for Applied Solar Energy. v.1, 1957- . 621.4705 JOU

REVIEWS & SURVEYS

K42. Advances in Applied Mechanics. Academic, v.1, 1948-
Supplements, 1961- . 531 Ad95

K43. Advances in Chemical Engineering. Academic, v.1, 1956.
660 Ad95

K44. Advances in Information Systems Science. Plenum, v.1,
1969- .

K45. Advances in Space Science and Technology. Academic,
v.1, 1959- . 629.13338 Ad95

K46. Progress in Nuclear Energy. Pergamon, v.1, 1956-
(Published in 12 series at irregular intervals) 539.5 P9463
  See: Wa p. 221.

See also the listings of reviews and surveys in the physical
science sections.

HISTORIES

See also GENERAL SCIENCE, No. A25-A35

General

K47. Davenport, W.H. & Rosenthal, D. comp. Engineering; its
Role and Function in Human Society. Pergamon, 1967. 620 D27e

K48. Kirby, R.S. et al. Engineering in History. McGraw-Hill,
1956. 620.9 K63en

K49. Oliver, J.W. History of American Technology. Ronald,
1956. 609.73 014h

K50. Singer, C. et al. History of Technology. Oxford, Clarendon
Pr., 1954-8. 5v. 609 Si6h
  See: Wa p. 43.

K51. Technology in Western Civilization. Oxford Univ. Pr.,
1967- . 609 T226
  See: Mech. Eng. 90:83-4, (Nov.) 1968.

K52. Technology and Culture. Wayne State Univ. Pr., v.1,
1959- 609.05 TEC

Special

K53. Aitchison, L. A History of Metals. Interscience, 1960.
2v. 669 Ai9h
  See: Wa p. 380.

K54. Burstall, A.F. A History of Mechanical Engineering.
London, Faber, 1963. (Reprinted, 1965) 621.09 B94h
  See: Wa p. 213-4

K55. Dünsheath, P. A History of Electrical Engineering. London,
Faber, 1962.
  See: Wa p. 229.

K56. Hewlett, R.G. & Anderson, O.E. A History of the United
States Atomic Energy Commission. Pa. State Univ. Pr., v.1,
1962-
  See: Wa p. 223.

K57. Kingery, R.A. et al. Men and Ideas in Engineering: Twelve
Histories from Illinois. Univ. of Ill. Pr., 1967. 620.72 K59m

K58. Smith, C.S. ed. Sources for the History of the Science of
Steel. 1532-1786. M.I.T. Pr., 1968.
  See: J. Chem. Doc. 8:248, 1968; Choice 5:995, 1968.

K59. Straub, H. A History of Civil Engineering: an Outline from
Ancient to Modern Times. London, Hill, 1952. (reprinted by
M.I.T. Pr., 1964) Translated from 1949 German edition. 690.9
St8gEr
  See: Wa p. 262.

K60. U.S. National Aeronautics and Space Administration. Astro-
nautics and Aeronautics; Chronology on Science, Technology, and
Policy. Wash., 1963- . (Monthly with annual volume) 629.4
Un363a

172

K61. U.S. National Bureau of Standards. Measures for Progress. Wash., G.P.O., 1966. 389 Un36m
 See: Chem. & Eng. News 46:76, (Sept. 19) 1966.

K62. Von Braun, W. & Ordway, F. III. History of Rocketry and Space Travel. T.Y. Crowell, 1966. 629.409 V39h
 See: Sky & Telescope 34:251-2, 1967.

K63. Wherry, J.H. Automobiles of the World: the Story of the Development of the Automobile with Rare Illustrations from a Score of Nations. Chilton, 1968.
 See: Lib. J. 94:533, 1969.

DICTIONARIES & ENCYCLOPEDIAS

General

K64. Carter, E.F. Dictionary of Inventions and Discoveries. Philosophical, 1967. 603 C24d
 See: New Tech. Bks. 52:88, 1967.

K65. Jones, F.D. & Schubert, P.B. Engineering Encyclopedia. 3d ed., N.Y., Industrial Pr., 1963. 620.3 J71e
 See: Wi-EI4.

K66. Modern Chinese — English Technical and General Dictionary. McGraw-Hill, 1963. 3v. 495.13 M73
 See: New Tech. Bks. 49:111, 1964. Wa p. 28.

K67. DeVries, L. & Herrman, T.M. English-German Technical and Engineering Dictionary. 2d ed., McGraw-Hill, 1967. 603 D49e
 See: New Tech. Bks. 53:201, 1968. Lib. J. 93:2639-40, 1968.

K68. DeVries, L. & Herrman, T.M. German-English Technical and Engineering Dictionary. 2d ed., McGraw-Hill, 1965. 603 D49g

K69. Walther, R. ed. Polytechnical Dictionary. Pergamon, 1968. 2v. v.1: English/German. v.2:German/English.

K70. Czerni, S. & Skrzyńska, M. eds. English-Polish and Polish-English Technological Dictionary. Macmillan, 1962. 2v.
 The authors have also edited: Shorter Technological Dictionary: Polish-English/English-Polish, Pergamon, 1963. (603 Sh81)
 See: Wa p. 27.

K71. Marolli, G. Dizionario Technico, Inglese-Italiano, Italiano-Inglese. 8th ed., Heinman, 1964.
  See: Lib. J. 89:3296, 1964, Wi-EA121. Wa p. 25.

K72. Robb, L.A. Engineers' Dictionary, Spanish-English, English-Spanish. 2d ed., Wiley, 1953. 620.3 R53e
  See: Wi-EA135. Wa p. 205.

Aeronautical & Astronautical Engineering

K73. Above and Beyond; The Encyclopedia of Aviation and Space Science. Chicago, New Horizons, 1968. 14v.
  See: Science 162:1259-60, 1968.

K74. Herrick, J.W. ed. Rocket Encyclopedia Illustrated. Aero Pub., 1959. 629.13338 H43r
  See: Sub. Bks. Bull. 56:253-4+, 1960. Wi-EI28. Wa p. 291 (note).

K75. McGraw-Hill Encyclopedia of Space. McGraw-Hill, 1968. 629.4 G76E.
  See: Sky & Telescope 36:255, 1968; Science 162:1259-60, 1968.

K76. McLaughlin, C. Space Age Dictionary. 2d ed., Van Nostrand, 1963. 629.403 M22s
  See: Sub. Bks. Bull. 56:173, 1959. Wi-EI29. Wa p. 291.

K77. Moser, R.C. Space-age Acronyms: Abbreviations and Designations. Plenum, 1964. 629.40148 M85s
  See: Wi-EI30. Wa p. 281.

K78. Roes, N. & Kennedy, W.E. The Space-Flight Encyclopedia. Follett, 1968.

K79. U.S. National Aeronautics and Space Administration. Dictionary of Technical Terms for Aerospace Use. Wash., G.P.O., 1965. 629.1303 Un375d

K80. Hyman, C.J. German-English, English-German Astronautics Dictionary. Plenum, 1967. 629.403 H99g

K81. Konarski, M.M. Russian-English Dictionary of Modern Terms in Aeronautics and Rocketry. Macmillan, 1962. 629.1303 K836r
  See: Sub. Bks Bull. 59:111+, 1963. Am. Scientist 51:192A-3A, 1963. Wa p. 285.

174

K82. Elsevier's Dictionary of Aeronautics in Six Languages:
English/American, French, Spanish, Italian, Portugese, and
German. N.Y., 1964. 629.1303 E176.
   See: Lib. J. 89:2592, 1964. Wi-EI33. Wa p. 283.

K83. North Atlantic Treaty Organization. AGARD Aeronautical
Multilingual Dictionary. Macmillan, 1960. Supplements, 1963-
(English, French, Spanish, German, Italian, Dutch, Turkish,
Russian. Supplement I adds Greek) 629.1303 N81a
   See: Lib. J. 89:1226-7, 1964. New Tech. Bks. 49:96, 1964.
Wi-EI34.

Agricultural Engineering

K84. Farrall, A.W. & Albrecht, C.F. ed. Agricultural Engineering: a Dictionary and Handbook. Interstate, 1965. 630.1503 F24a
   See: Science Books 1:109, 1965.

Chemical & Industrial Engineering

K84a. Dictionary of Chemical Engineering. Elsevier, 1969. 2v.
   See: Chem. & Eng. News 47:64, (Apr. 21) 1969.

K85. Encyclopedia of Engineering Materials and Processes.
Reinhold, 1963. 620.11 En192
   See: Chem. & Eng. News 42:44, (Jan. 6) 1964. Wi-EI147.

K86. Industrial and Engineering Chemistry. Modern Chemical
Processes. Reinhold, 1952-63. 7v. q660 In3m

K87. Kent, J.A. ed. Riegel's Industrial Chemistry. Reinhold,
1962. 660 R44i

K88. Mead, W.J. ed. Encyclopedia of Chemical Process Equipment. Reinhold, 1964. 660.28303 En19
   See: Wi-ED26. Wa p. 355 (note).

K89. Modern Chemical Engineering. Reinhold, 1963-   (a
monographic series)
   See: Chem. & Eng. News 41:114-5, 1963.

K90. Thorpe, J.F. & Whiteley, M.A. Thorpe's Dictionary of
Applied Chemistry. 4th ed., Longmans, 1937-56. 12v. 540.3 T39d
   See: Wi-ED29. Wa p. 352.

K91. Ullmann, F. Ullmanns Encyklopädie der technischen Chemie.
3d ed., Berlin, Urban, v.1,1951- . 660.3 Ul4e
   See: Wi-ED30. Wa p. 352.

Civil Engineering

K92. Comrie, J. ed. Civil Engineering Reference Book. 2d ed.,
London, Butterworths, 1961. 4v. 620.2 P94c
   See: Wi-EI70. Wa p. 261.

K93. Gros, E. comp. Constructional Engineering Dictionary:
Russian, English, French, German. London, Scientific Information
Consultants Ltd., 1965. 690.3 G911r

Electrical & Electronic Engineering

K94. Carter, H. Dictionary of Electronics. 3d ed., Hart, 1967.
621.38103 C245d
   See: Lib. J. 92:4139, 1967; Sp. Libs. 57:417, 1966.

K95. Graf, R.F. Modern Dictionary of Electronics. 3d ed., Sams,
1968. 621.38103 M72
   See: Sp. Libs. 59:119, 1968.

K96. Institute of Radio Engineers. IRE Dictionary of Electronic
Terms and Symbols Compiled from IRE Standards. N.Y., 1961.
621.3803 In7i
   See: Wa p. 234.

K97. Markus, J. Electronics and Nucleonics Dictionary. 3d ed.,
McGraw-Hill, 1966. (title varies) 621.38103 C77
   See: New Tech. Bks. 52:60, 1967. Lib. J. 93:1585, 1968.

K98. Simonyi, K. Foundations of Electrical Engineering.
Macmillan, 1963- . 621.3 Si5fE

K99. Susskind, C. ed. The Encyclopedia of Electronics.
Reinhold, 1962. 621.38103 Su8e
   See: Wi-EI97. Wa p. 233.

K100. Hohn, E. Dictionary of Electrotechnology: German-
English. Barnes & Noble, 1967. 621.303 H68d
   See: Lib. J. 92:1608, 1967.

176

K101. Hyman, C. L. German-English, English-German Electronics Dictionary. Plenum, 1965. 621.38103 H98g
  See: Sp. Libs. 57:67, 1966.

K102. Schwenkhagen, H. F. Dictionary of Electrical Engineering: German-English, English-German. Wiley, 1963. 621.303 Sch 9fE
  See: Wa p. 228 (note).

K103. Dorian, A. F. comp. Six-language Dictionary of Electronics, Automation and Scientific Instruments; a Comprehensive Dictionary in English, French, German, Italian, Spanish, and Russian. Prentice-Hall, 1962. 621.3803 Si97
  See: New Tech. Bks. 48:139, 1963.

Mechanical Engineering

K104. Georgano, G. N., ed. The Complete Encyclopedia of Motorcars, 1885-1968. Dutton, 1968.
  See: RQ 8:139-40, 1969. Booklist 65:524, 1969.

K105. Horner, J. G. A Dictionary of Mechanical Engineering Terms. 9th ed., London, Technical Pr., 1967. 621.03 H781
  See: Wi-EI139. Wa p. 211.

K106. Nayler, J. L. & Nayler, G. H. F. Dictionary of Mechanical Engineering. Hart, 1967. 621.03 N23d
  See: Choice 5:182, 1968.

Military & Naval Engineering

K107. Craig, H. comp. A Bibliography of Encyclopedias and Dictionaries Dealing with Military, Naval and Maritime Affairs, 1577-1961. 2d ed., Fondren Library, Rice University, 1962. 016.35503 C84b
  See: Wa p. 254.

K108. U. S Division of Naval History. Dictionary of American Naval Fighting Ships. Wash., G. P. O., v. 1, 1959-   . 359.32 Un35d
  See: Wi-CI214.

177

K109. U.S. Joint Chiefs of Staff. A Dictionary of United States Military Terms, Prepared for Joint Usage of the Armed Services. Wash., Public Affairs Pr., 1959/69-   . 355 Un375j
  See: Wi-CI206. Wa p. 254.

K110. Elsevier's Nautical Dictionary. Elsevier, 1965-6. 3v. 623.803 E176
  See: Lib. J. 91:4086, 1966; 92:1143-4, 1967.

K111. Kerchove, R. de. International Maritime Dictionary; an Encyclopedic Dictionary of Useful Maritime Terms and Phrases, Together with Equivalents in French and German. 2d ed., Van Nostrand, 1961. 387.03 K45i

Mining & Metallurgical Engineering

K112. Merriman, A.D. A Concise Encyclopedia of Metallurgy. Elsevier, 1965. (Based on the author's Dictionary of Metallurgy, 1958) 669.03 M55d
  See: New Tech. Bks. 50:353, 1965.

K113. Nelson, A. Dictionary of Mining. Philosophical, 1965. 622.03 N33d

K114. Pryor, E.J. Dictionary of Mineral Technology. London, Mining, 1963. 662.03 P956d
  See: Wa p. 378.

K115. Sittig, M. ed. Inorganic Chemical and Metallurgical Process Encyclopedia. Noyes, 1968.

K116. Clason, W.E. comp. Elsevier's Dictionary of Metallurgy in Six Languages: English/American, French, Spanish, Italian, Dutch, and German. Elsevier, 1967. 669.03 E176
  See: New Tech. Bks. 52:211, 1967.

Nuclear Engineering

K117. National Research Council Conference on Glossary of Terms in Nuclear Science and Technology. A Glossary of Terms in Nuclear Science and Technology. N.Y., American Society of Mechanical Engineers, 1957. 539.03 N21g
  See: Wi-EI204. Wa p. 220.

K118. Weinstein, R. et al. Nuclear Engineering Fundamentals. McGraw-Hill, 1964. 621.48 W433n

Photography & Optical Engineering

K119. American Society of Photogrammetry. Manual of Color Aerial Photography. Falls Church, Va., 1968. 778.35 Am35m

K 120. Focal Encyclopedia of Photography. 2d ed., N.Y., Focal Pr., 1965. 2v. 770.3 F682
  See: **Sub. Bks. Bull.** 55:35, 1958. Science 154:639, 1966. **Wi-BF78.**

K121. Kingslake, R. ed. Applied Optics and Optical Engineering. Academic, 1965- . 621.32 K61a
  See: Sky & Telescope 36:38-40, 1968.

K122. Elsevier's Dictionary of Photography. (English-French-German) N.Y., 1965. 770.3 E176.

HANDBOOKS & YEARBOOKS

General

K123. Hawkins, G.A. ed. Student's Engineering Manual. McGraw-Hill, 1968.

K124. Liebers, A. The Engineers Handbook. Key, 1968.

K125. Perry, R.H. Engineering Manual; a Practical Reference of Data and Methods in Architectural, Chemical, Civil, Electrical, Mechanical, and Nuclear Engineering. 2d ed., McGraw-Hill, 1967. 620.2 P42e
  See: New Tech. Bks. 52:268, 1967; Sp. Libs. 58:733, 1967.

K126. Potter, J.H. Handbook of Engineering Sciences. Van Nostrand, 1967. 2v. 620.2 P853e
  See: Mech. Eng. 90:84-5, (June) 1968.

K127. Souders, M. The Engineer's Companion: a Concise Handbook of Engineering Fundamentals. Wiley, 1966. 621.002 So8e
  See: New Tech. Bks. 51:91, 1966.

K128. Yakovlev, K. P. ed. Handbook for Engineers. (Translated from the Russian by G. O. Harding) Pergamon, 1965-
620 Ia 5KEh
  See: New Tech. Bks. 50:268, 1965.

Aeronautical & Astronautical Engineering

See also the unit of Military & Naval Engineering (Nos. K107-K111)

K129. Aerospace Facts and Figures. Aero Pub., 1945-   (Irregular, 1945-53; annual, 1955-   .) 629.13 Av31
  See: New Tech. Bks. 49:70, 1964.

K130. Aerospace Yearbook. Wash., American Aviation Pub.,
1919-   . (title varies) 629.13 Ai74.
  See: Wi-EI45. Wa p. 286.

K131. Allen, R. Great Airports of the World. New Rochelle, N. Y.,
Sportshelf, 1964.

K132. Corliss, W. R. Scientific Satellites. Wash., G. P. O., 1967
(NASA SP-133). 629.434 C81s
  See: Physics Today 21:91-3, (Sept.) 1968; Sky & Telescope
35:247, 1968.

K133. Haviland, R. P. & House, C. M. Handbook of Satellites and
Space Vehicles. Van Nostrand, 1965. 629.4 H29h
  See: Science 149:173, 1965. Wi-1E16.

K134. Johnson, F. S. ed. Satellite Environment Handbook. 2d ed.,
Stanford Univ. Pr., 1965. 551.514 J33s
  See: Science 148:814, 1965. Wi-1E17.

K135. Kit, B. & Evered, D. S. Rocket Propellant Handbook.
Macmillan, 1960. 629.13354 K64r
  See: Sub. Bks. Bull. 56:619, 1960.

K136. Lockheed Aircraft Corporation. Space Materials Handbook.
Addison Wesley, 1966. 629.4 L81s
  See: Lib. J. 91:1153, 1966.

K137. McEntee, H. G. The Model Aircraft Handbook. Crowell,
1968.
  See: New Tech. Bks. 54:61, 1969.

K138. Morgan, L. Airliners of the World. Aero Pub., 1967.
629.13334 M82a
  See: Booklist 63:1024, 1967.

K139. Syracuse University Research Institute. Aerospace Structural Metals Handbook. Syracuse Univ. Pr., 1963. 2v. (Looseleaf)
q620.16 Sy81a

K140. Webb, P. ed. Bioastronautics Data Book. Wash., G.P.O.,
1964. 616.98 W383b
  See: Science 148:1457, 1965.

K141. American Aviation. Wash., American Aviation Pubs., v.1,
1937- . (Frequency and title vary) Absorbed Aerospace Technology
(Title varies) in June 1968. 629.105 AME

Agricultural Engineering

K142. Richey, C.B. et al. Agricultural Engineer's Handbook.
McGraw-Hill, 1961. 630 R39a
  See: Wi-EK35.

Chemical & Industrial Engineering

K143. American Gas Association. Gas Enginers Handbook; Fuel
Gas Engineering Practices. Industrial, 1965. 665.7 Am31ga
  See: New Tech. Bks. 51:139, 1966.

K144. Chemical Engineers' Handbook. 4th ed., McGraw-Hill,
1963. 660 P42c
  See: New Tech. Bks. 49:33, 1964. Wi-EI60.

K145. Chemical Engineering Practice. London, Butterworths,
1956- . 660 C86c
  See: Wi-EI61. Wa p. 355.

K146. Considine, D.M. & Ross, S.D. Handbook of Applied Instrumentation. McGraw-Hill, 1964. 620 C768h
  See: Wa p. 412.

K147. Handbook of Industrial Research Management. 2d ed.,
Reinhold, 1968. 658.57 H51h

K148. Petroleum Processing Handbook. McGraw-Hill, 1967.
665.53 P448

K149. Society of the Plastics Industry. SPI Plastics Engineering Handbook. 3d ed., Reinhold, 1960. 668.41 Sols
  See: Wi-EI67. Wa p. 407

K150. System Engineering Handbook. McGraw-Hill, 1965.

Civil Engineering

K151. Abbott, R.W. American Civil Engineering Practice. Wiley, 1956-7. 3v. (Successor to Merriam's American Civil Engineers' Handbook) 620 Ab3a
  See: A.L.A. Booklist 53:239, 1957. Wi-EI69.

K152. Gaylord, E.H., Jr. & Gaylord, C.N. Structural Engineering Handbook. McGraw-Hill, 1968. 624.1 G25s

K153. Merritt, F.S. ed. Standard Handbook for Civil Engineers. McGraw-Hill, 1968.

K154. Seelye, E.E. Data Book for Civil Engineers. 3d ed., Wiley, 1957-60. 2v. (v.1: Designs; v.2: Specifications and costs) q620 Se3d
  See: Sub. Bks. Bull. 54:171, 1957. Wa p. 262.

Electrical & Electronic Engineering

General

K155. American Electricians' Handbook: a Reference Book for Practical Electrical Workers. 8th ed., McGraw-Hill, 1961. 621.3 C874a

K156. Clifford, M. Electronics Data Handbook, N.Y., Gernsback, 1964. 621.381 C61e
  See: New Tech. Bks. 49:317, 1964.

K157. Crowhurst, N.H. Electronics Reference Databook. TAB Bks., 1969.

K158. Handbook of Electronic Engineering. Chemical Rubber Co., 1967. 621.36 H87e

K159. Illuminating Engineering Society. IES Lighting Handbook; the Standard Lighting Guide. 4th ed., N.Y., 1966. q621.32 I161i

K160. Standard Handbook for Electrical Engineers. 10th ed.,
McGraw-Hill, 1968. 621.3 St271

K161. Stetka, F. & Brandon, M.M. NFPA Handbook of the National
Electrical Code. McGraw-Hill, 1966. (Replaces the National
Electrical Code Handbook, 11th ed., 1963) 621.3007 N213
  See: New Tech. Bks. 52:90-1, 1967. Wi-EI111.

K162. U.S. Department of Commerce. Bureau of International
Commerce. Electrical Current Abroad. Wash., 1959-   .
621.3 Un333e
  See: Wa p. 229.

        Radio, Television, Stereo

K163. American Radio Relay League. Radio Amateur's Handbook.
1st ed., 1926-   . (Annual; 46th ed., 1969) 621.365 Am3r
  See: Wi-EI121. Wa p. 239 (note).

K164. Boyce, W.F. Hi-fi Stereo Handbook. 3d ed., Sams, 1967.
  See: New Tech. Bks. 57:271, 1967.

K165. Hicks, D.E. Citizens Band Radio Handbook. 3d ed., Sams,
1967. (Title varies) 621.3846 H52c
  See: New Tech. Bks. 52:348, 1968.

K166. Jones, V.A. North American Radio-TV Station Guide.
5th ed., Sams, 1968. 621.38416 J72n
  See: New Tech. Bks. 54:29, 1969.

K167. Pyle, H.S. General Class Amateur License Handbook.
2d ed., Sams, 1968. (title varies)
  See: New Tech. Bks. 53:166-7, 1968.

K168. Radio Handbook. 17th ed., Summerland, Calif., Editors
& Engineers, 1967. 621.384 R118

K169. Rider, J.F. Perpetual Trouble Shooter's Manual. N.Y.,
v.1, 1933-   . (Looseleaf) 621.365 R43p
  See: Wi-EI124.

K170. Rider, J.F. Television Manual. N.Y., v.1, 1948-   .
(looseleaf) q621.388 R43t
  See: Wi-EI125.

K171. Westman, H.P. et al. Reference Data for Radio Engineers.
5th ed., Sams, 1968.

Mechanical Engineering

K172. American Society of Mechanical Engineers. ASME Handbook.
2d ed., N.Y., 1964- . 620.18 Am34a
  See: Wi-EI143, 1EI34. Wa p. 212-3.

K173. ASHRAE Guide and Data Book. N.Y., American Society of
Heating, Refrigerating and Air-conditioning Engineers, 1963-64.
2v. 697 Am35a
  See: Wi-EI126.

K174. Chilton's Auto Repair Manual; American Cars from 1961 to
1969 - . Plus Volkswagens. Chilton, 1967. q656.3 C44
  See: New Tech. Bks. 53:25, 1968.

K175. Glenn's Foreign Car Repair Manual. 2d ed., Chilton, 1966.
(1st ed., 1963)

K176. Jane's World Railways. 9th ed., McGraw-Hill, 1966.
  See: New Tech. Bks. 52:62-3, 1967; Sci. Am. 216:142-3,
(Jan.) 1967.

K177. Machinery's Handbook. Industrial Pr., 1st ed.,1914-   .
(18th ed., 1968) 621 M18m
  See: Wi-EI152.

K178. SAE Handbook. N.Y., Society of Automotive Engineers,
1926- . (Annual) 656.3 So12s

K179. Standard Handbook for Mechanical Engineers. McGraw-Hill,
1st ed., 1916- . (7th ed., 1967. First five editions edited by
L.S. Marks) 620.8 M34m
  See: Choice 5:182-3, 1968.

K180. Strock, C. & Koral, R.R. ed. Handbook of Airconditioning,
Heating and Ventilating. 2d ed., Industrial Pr., 1965.
  See: New Tech. Bks. 51:34, 1966.

Military & Naval Engineering

K181. Jane's All the World's Aircraft. McGraw-Hill, v.1, 1909-
Annual. (title varies) 629.105 AL
  See: Wi-EI46. Wa p. 288.

K182. Jane's Fighting Ships. McGraw-Hill, v. 1, 1898- .
Annual. (title varies) 359 F46
  See: Wi-EI162. Wa p. 259-60.

K183. Smith, J.E. ed. Small Arms of the World. 8th ed.,
Stackpole, 1966.
  See: Sp. Libs. 58:274, 1967.

K184. U.S. Aircraft, Missiles, and Spacecraft. National Aerospace
Education Council, 1957- . Annual.

K185. U.S. Bureau of Customs. Merchant Vessels of the United
States. 1st ed., 1866- . Annual. 387 Un35
  See: Wi-CH220.

Mining & Metallurgical Engineering

K186. American Society for Metals. Metals Handbook. Novelty,
Ohio, 1st ed., 1927- . (5v. projected for 8th ed. 1961- .)
669.1 Am35m
  See: New Tech. Bks. 50:97, 1965. Wi-EI191. Wa p. 379.

K187. American Society of Tool & Manufacturing Engineers. Die
Design Handbook: a Practical Reference Book on Process Analysis,
Product Design, Metal Movements, Materials, and Proved Die
Designs for Every Class of Sheet Metal Presswork. 2d ed.,
McGraw-Hill, 1966. 621.8 Am32d

K188. American Welding Society. Welding Handbook. 5th ed.,
N.Y., 1962-7. 5v. 669.173 Am31we

K189. Hampel, C.A., ed. Rare Metals Handbook. 2d ed.,
Reinhold, 1961. 669.7 H18r

K190. Parker, E.R. Materials Data Book. McGraw-Hill, 1967.
  See: New Tech. Bks. 53:32, 1968.

K191. Pearson, W.B. A Handbook of Lattice Spacings and Structures of Metals and Alloys. Pergamon, 1958-67. 2v. 669 M562
  See: New Tech. Bks. 53:99, 1968.

K192. Smithells, C.J., ed. Metals Reference Book. 4th ed.,
Plenum, 1967. 3v. 669 Sm6m
  See: Wi-EI193.

K193. Woldman, N.E. Engineering Alloys: Names, Properties,
Uses. 4th ed., Cleveland, American Society for Metals, 1962.
669 W83e
  See: Science 144:103, 1963. Wi-EI197. Wa p. 380.

Nuclear Engineering

See also the section on PHYSICS

K194. Etherington, H. ed. Nuclear Engineering Handbook.
McGraw-Hill, 1958. 539.5 Et3n
  See: Science 129:1137, 1959. Wi-EI209. Wa p. 222.

K195. Plutonium Handbook; a Guide to the Technology. Gordon
& Breach, 1967- . 546.434 P746
  See: New Tech. Bks. 53:33, 1968.

K196. Reactor Handbook. 2d ed., Interscience, 1960-4. 4v.
539.3 Un3rea
  See: New Tech. Bks. 46:41, 1961. Wi-EI211.

INFORMATION ACTIVITIES

**K197. American Institute of Chemical Engineers. Chemical
Engineering Thesaurus. N.Y., 1961. 660.78 Am35c**

K198. American Society for Metals. ASM Thesaurus of Metallurgical
Terms. **1968. 025.36 Am3a**

K199. Battelle Memorial **Institute.** A Survey of Science-Informa-
tion Manpower in Engineering and the Natural Sciences. Final
Report to the National Science Foundation Columbus, Ohio,
November 30, 1966.
  See: Sp. Libs. 58:599, 1967.

K200. Engineering Societies Library. Bibliography on Filing,
Classification and Indexing Systems, and Thesauri for Engineering
Offices. N.Y., 1966. 016.62 N485f

  See also: Engineers Joint Council. Thesaurus of Scientific and
Engineering Terms. (NO. A204)

STANDARDS & SPECIFICATIONS

K201. American Society for Testing and Materials. Book of ASTM
Standards. 1st ed., 1910- . 32 parts. Annual. (Title, number
of parts and frequency vary) 620.6 AMAS
  See: Wi-EA204.

186

K202. American Society for Testing and Materials. Index to ASTM Standards. v. 1, 1928-   . Annual. 620.6 AMASA

K203. American Standards Association. American Standards. N.Y., 1923-   . 620 Am33a

K204. American Standards Association. Catalog of American Standards. N.Y., 1923-   . A620.1 Am36a

K205. British Standards Institution. Yearbook. London, 1937-   . Annual. 672 B77h
  See: Wi-EA206.

K206. Struglia, E.J. Standards and Specifications Information Sources; a Guide to Literature and to Public and Private Agencies Concerned with Technological Uniformities. Gale, 1965. 016.3896 St8s
  See: Sp. Libs. 57:260-1, 1966.

K207. U.S. General Services Administration. Federal Standardization. Wash., G.P.O., 1965. 658.516 Un31f

K208. U.S. National Bureau of Standards. Publications of the National Bureau of Standards, 1901 to June 30, 1947. Wash., 1948. Supplementary lists, 1958-   . 016.389 Un32p
  See: Wi-EA207.

BIOGRAPHIES & DIRECTORIES

K209. Directory of Engineering Societies and Related Organizations. N.Y., Engineers Joint Council, 1968. (Formerly Engineering Societies Directory, 1948-   . 620.2 ENGD

K210. National Society of Professional Engineers. Directory of Engineers in Private Practice. Wash., 1965-   . Annual. (title varies) 620.2 NATS
  See: Mech. Eng. 90:79, (Oct.) 1968.

K211. U.S. National Aeronautics and Space Administration. Space Scientists and Engineers: Selected Biographical and Bibliographical Listing, 1957-61. Wash., G.P.O., 1962. (NASA Sp-5) 629.4 Un36n no. 5
  See: Wi-EI47. Wa p. 294.

K212. Who's Who in Engineering; a Biographical Dictionary of the Engineering Profession. N.Y., Lewis Historical Pub., 1st ed., 1922- . (9th ed., 1964) 926.2 W622
  See: Wi-EA197. Wa p. 207.

K213. Who's Who of British Engineers, 1968. Ohio Univ. Pr., 1969.
  See: Science News 95:236, 1969.

K214. World Aviation Directory. Wash., American Aviation Pub., v.1, 1940- . Semiannual. 629.13 Am351
  See: Wi-EI37.

K215. World Space Directory. Wash., American Aviation Pub., v.1, 1963- . Semiannual. 629.4 W89
  See: Wi-EI38. Wa p. 292.

SERIALS

K216. Massachusetts Institute of Technology Library. Current Serials and Journals in the M.I.T. Libraries. 1st ed., 1957- 016.505 M38c

K217. U.S. Library of Congress. Aeronautical and Space Serial Publications: a World List. Wash., 1962. 016.6291305 Un3c
  See: Wi-EI18. Wa p. 285.

CURRENT INFORMATION: SELECTION AIDS

  See:
    Fetros, J.G. How-To-Do-It Books: Building the Collection. Lib. J. 94:1595-7, 1969.

Consult current issues of such periodicals as:

    AIAA Bulletin, v.1, 1964- .
    Astronautics & Aeronautics, v.1, 1957- .
    Chemical & Engineering News, v.1, 1923-
    Civil Engineering, v.1, 1930- .
    Electronics, v.1, 1930- . (including Annual Buyers' Guide issue)
    Electronics World, v.1, 1919- .
    Engineer, v.1, 1956- . (including Buyers Guide)
    Engineering News-Record, v.1, 1874- .
    IEEE Spectrum, v.1, 1964- .

Industrial & Engineering Chemistry, v. 1, 1909-
Mechanical Engineering, v. 1, 1906-   .
Modern Plastics, v. 1, 1925-   (including Modern Plastics
   Encyclopedia)
Popular Electronics, v. 1, 1954-
Popular Science Monthly. v. 1, 1872-   .
QST, v. 1, 1915-   .
Radio and Television Weekly, v. 1, 1916-
Railway Age, v. 1, 1856-
Steel, v. 1, 1882-

INDEX

190

Advances in Internal Medicine, H34.
Advances in Macromolecular Chemistry, D36.
Advances in Mathematics, B13.
Advances in Nuclear Physics, C20.
Advances in Organic Chemistry, D37.
Advances in Physical Organic Chemistry, D38.
Advances in Plasma Physics, C21.
Advances in Polymer Science, D39.
Advances in Quantum Chemistry, D40.
Advances in Space Science and Technology, K45.
Advances in Theoretical Physics, C22.
Advances in Veterinary Sciences, J93.
Aerospace Facts and Figures, K129.
Aerospace Medicine and Biology, H22.
Aerospace Yearbook, K130.
Agricultural Index, J6.
Air University Library Index to Military Periodicals, K17.
Aitchison. History of Metals, K53.
Aleksandrov et al. Mathematics, B27.
Alexander. Birds of the Ocean, G220.
Allbritton. The Fabric of Geology, F40.
Allen. Astrophysical Quantities, E42.
Allen. Great Airports of the World, K131.
Allen. Star-names and Their Meanings, E43.
Almanac of the Canning, Freezing, Preserving Industries, J112.
Alt. Plans for a National Physics Information System, C71.
Alter. Lunar Atlas, E64.
Alter. Pictorial Guide to the Moon, E65.
Altman & Dittmer. Biological Data Book, G49.
Altman & Dittmer. Metabolism; Biological Handbook, G75.
Altsheler. Natural History Index-guide, G5.
American Aviation, K141.
Am. Chem. Soc. Chemical Nomenclature, D153.
Am. Chem. Soc. Directory of Graduate Research, D11.
Am. Chem. Soc. Key to Pharmaceutical and Medicinal Chemistry, D10.
Am. Chem. Soc. Literature of Chemical Technology, D7.
Am. Chem. Soc. Literature Resources for Chemical Processing Industries, D9.
Am. Chem. Soc. Reagent Chemicals, D162.
Am. Chem. Soc. Searching the Chemical Literature, D8.
American Dental Directory. H131.
American Electricians' Handbook, K155.
American Gas Assn. Gas Engineers Handbook, K143.

American Geological Institute. Glossary of Geology, F41.

American Home Economics Assn. Textile Handbook, J128.

American Institute of Biological Sciences. Biology Teachers' Handbook, G50.

American Institute of Biological Sciences. Directory of Bioscience Departments, G69.

American Institute of Chemical Engineers. Chemical Engineering Thesaurus, K197.

American Institute of Physics. **Glossary of Terms** Frequently Used in Radio Astronomy, E22.

American Institute of Physics Handbook, **C58.**

American Journal of Botany, G89.

American Kennel Club. The Complete Dog **Book,** J97.

American Mathematical Society. Contents of Contemporary Mathematical Journals, B5.

American Medical Assn. Current Medical Terminology, H52.

American Medical Assn. Distribution of Physicians, Hospitals, and Hospital Beds, H123.

American Medical Assn. Journal, H20.

American Medical Assn. New Drugs, H105.

American Medical Assn. Today's Health Guide, H73.

American Medical Directory, H132.

American Medical Writer's Assn. Directory of Free-lance Writers, Editors & Researchers, A139.

American Men of Medicine, H127.

American Men of Science, A87.

American Meteorological Society. Compendium of Meteorology, F101.

American Pharmaceutical Assn. The National Formulary, H106.

American Physiological Society. Handbook of Physiology, H87.

American Public Health Assn. Control of Communicable Diseases in Man, H114.

American Radio Relay League. Radio Amateur's Handbook, K163.

American Scientific Books, A1.

American Society for Metals. ASM Thesaurus, K198.

American Society for Metals. Metals Handbook, K186.

American Society for Testing & Materials. Book of ASTM Standards, K201.

American Society for Testing & Materials. CODEN for Periodical Titles, A190.

American Society for Testing & Materials. Index to ASTM Standards, K202.

American Society of Mechanical Engineers. ASME Handbook, K172.

American Society of Photogrammetry. Manual of Color Aerial Photography, K119.

American Society of Tool & Manufacturing Engineers. Die Design Handbook, K187.

American Standards Assn. American Standards, K203.

American Standards Assn. American Standards for Periodical Title Abbreviations, A145.

American Standards Assn. Catalog of American Standards, K204.

American Translators Assn. ATA Professional Services Directory, A154.

American Veterinary Medical Assn. Directory, J105.

American Welding Society. Welding Handbook, K188.

Analytical Chemistry, U.S.S.R., D94.

Andrews. Index of Generic Names of Fossil Plants, F157.

Andrews. World List of Pharmacy Periodicals, H142.

Animal Breeding Abstracts, J21.

Animal Health Yearbook, J98.

Animal Kingdom, G207.

Annals of Mathematics, B14.

Annals of the International Geophysical Year, F26.

Annals of the IQSY, F102.

Annee Biologique, G14.

Annotated Bibliography of Economic Geology, F8.

Annotated Bibliography on Hydrology and Sedimentation, F87.

Annual Reports on the Progress in Chemistry, D41.

Annual Review of Astronomy & Astrophysics, E7.

Annual Review of Biochemistry, D42.

Annual Review of Entomology, G155.

Annual Review of Genetics, G163.

Annual Review of Information Science, A202.

Annual Review of Medicine, H35.

Annual Review of Microbiology, G171.

Annual Review of Nuclear Science, C23.

Annual Review of Physical Chemistry, D43.

Anthony. Sources of Information on Atomic Energy, C1.

APCA Abstracts, K18.

Applied Mechanics Reviews, K19.

Applied Science & Technology Index, K11.

Arber. Herbals, Their Origin and Evolution, G99.

Arnell. Standard Graphical Symbols, A146.

Artschwager. Dictionary of Biological Equivalents, G33.

Ashburn. Laser Literature, K20.

ASHRAE Guide & Data Book, K173.

Asimov. Understanding Physics, C47.

Aslib Directory, A100.

Assn. of American Medical Colleges. Medical School Admission Requirements, H133.

Assn. of Official Agricultural Chemists. Official Methods of Analysis, J46.

Astronomischer Jahresbericht, E1.

Astrophysical Journal, E8.

Atomic Handbook, C73.
Audubon Nature Encyclopedia, G39.
Auerbach Computer Notebook, B68.
Auerbach. Geschichtstafeln der Physik, C29.
Australian Science Index, A22.
Axelrod & Schultz. Handbook of Tropical Aquarium Fishes, G230.

Bacteriological Reviews, G172.
Baker. Bibliography of Food, J108.
Baker. Dictionary of Mathematics, B28.
Ballentyne & Walker. Dictionary of Named Effects & Lws, A55.
Bancroft. Topics in Intermediate Statistical Methods, B81.
Barabashov et al. Atlas of the Moon's Far Side, E66.
Barnes & Etherington. Drug Dosage in Laboratory Animals, J99.
Barnhart. Biographical Notes Upon Botanists, G143.
Barton & Barton. Guide to the Constellations, E50.
BASIC, G7.
Bates. Scientific Societies in the U.S., A122.
Battelle Memorial Institute. Directory of Selected Scientific
  Institutions in the USSR, A112.
Battelle Memorial Institute. Survey of Science-information
  Manpower, K199.
Battelle Technical Review, K21.
Bauer et al. Bray's Clinical Laboratory Methods, H116.
Bäumler. Century of Chemistry, D49.
Beaton & McHenry. Nutrition, H88.
Bedevian. Illustrated Polyglottic Dictionary of Plant Names, G112.
Beeson & McDermott. Cecil-Loeb Textbook of Medicine, H115.
Beilstein's Handbuch der organischen Chemie, D111.
Bender. Dictionary of Nutrition and Food Technology, J113.
Bender. Dietetic Foods, J114.
Bennett. Chemical Formulary, D158.
Bennett. Concise Chemical and Technical Dictionary, D62.
Bent. Life Histories of North American Birds, G221.
Berichte über die gesamte Biologie, G6.
Berry. Short History of Astronomy, E13.
Besançon. The Encyclopedia of Physics, C48.
Besterman. World Bibliography of Bibliographies, A11.
Better Homes & Gardens Family Medical Guide, H74.
Beyer. CRC Handbook of Tables for Probability and Statistics,
  B82.
Bibliographia Oceanographica, F138.
Bibliographical Current List of Papers, Reports and Proceedings
  of International Meetings, A134.

Bibliographie des Sciences de la Terre, F9.
Bibliographie mensuelle de l'Astronomie, E3.
Bibliography and Index of Geology, F10.
Bibliography and Index of Geology Exclusive of North America, F10.
Bibliography of Fossil Vertebrates, F160.
Bibliography of Natural Radio Emission from Astronomical Sources, E2.
Bibliography of North American Geology, F11.
Bibliography of Proceedings of International Meetings, A135.
Bibliography of Seismology, F174.
Bibliography of Soil Science, Fertilizers and General Agronomy, J22.
Bibliography of the History of Medicine, H41.
Bibliotheca Historico-naturalis, G187.
Bibliotheca Zoologica I., G188.
Bibliotheca Zoologica II., G189.
Bidwell. History of Agriculture in the Northern U.S., J37.
Biographical Dictionary of Scientists, A87a.
Biographisches Lexikon der hervorragenden Ärzte, H7.
Biographisches Lexikon der hervorragenden Ärzte der letzten Jahre, H8.
Biological Abstracts, G7.
Biological & Agricultural Index, G9.
Biomedical Sciences Instrumentation, G44.
BioResearch Index, G8.
BioScience, G13.
Birch. Maps, Topographical & Statistical, F69.
Bishop. Handbook of Salamanders, G245.
Bittar & Bittar. The Biological Basis of Medicine, G45.
Blackwelder. Taxonomy, G203.
Blair et al. Vertebrates of the U.S., G208.
Blaisdell et al. Sources of Information in Transportation, K3.
Blake. Geographical Guide to Floras of the World, G84.
Blake & Roos. Medical Reference Works, H1.
Blanchard & Ostvold. Literature of Agricultural Research, J1.
Blinkov & Glezer. The Human Brain in Figures & Tables, H34.
Bliss. Statistics in Biology, G59.
Blum. Concise Russian-English Scientific Dictionary, A71.
Blyth. Geological Maps and Their Interpretation, F70.
Bolton. Catalogue of Scientific & Technical Periodicals, A191.
Bolton. Select Bibliography of Chemistry, D12.
Book of Popular Science, A42.
Borgstrom. Principles of Food Science, J115.
Börner. Minerals, Rocks, and Gemstones, F121.
Borrer. Dictionary of Word Roots and Combining Forms, G36.
Boss. General Catalogue of 33,342 Stars, E51.
Botanical Abstracts, G81.
Botanical Society of America. Fifty Years of Botany, G100.

Botanisches Centralblatt, G82.

Bottle. Use of Chemical Literature, D1.

Bottle & Wyatt. Use of Biological Literature, G1.

Bouliére. Elements d'un Guide bibliographique du Naturaliste, G2.

Bowditch. American Practical Navigator, E38.

Bowker Associates. U.S. Patent Previews, A179.

Boyce. Hi-Fi Stereo Handbook, K164.

Boyd. Fundamentals of Immunology, H90.

Boyd. Pathology for the Physician, H91.

Boyer. History of Mathematics, B16.

Brachet & Mirsky. The Cell, G46.

Bradbury. The Evolution of the Microscope, G21.

Bradley & Barnes. Chinese-English Glossary of Mineral Names, F120.

Bradshaw. PB-AD Report Index, A172.

Brady. Materials Handbook, A77.

Bragg & Claringbull. Crystal Structures of Minerals, F122.

Bray. Russian-English Scientific & Technical Dictionary, A72.

Bray's Clinical Laboratory Methods, H116.

Brazol. Dictionary of Meteorological & Related Terms, F105.

Britannica Yearbook of Science and the Future, A51.

British Abstracts, D13.

British Museum (Natural History) Library. Catalogue, G10.

British Standards Institution. Yearbook, K205.

British Technology Index, K12.

Brooklyn Botanic Garden. Origins of American Horticulture, J32.

Building Science Abstracts, K22.

Bulletin des Sciences mathématiques, B6.

Bulletin Signalétique, A18.

Bulloch. The History of Bacteriology, G176.

Burgunker. Russian-English Dictionary of Earth Sciences, F51.

Burhop. High Energy Physics, C53.

Burington. Handbook of Mathematical Tables & Formulas, B51.

Burkett & Plumb. How to Find Out in Electrical Engineering, K4.

Burlak. Russian-English Mathematical Vocabulary, B42.

Burman. How to Find Out in Chemistry, D2.

Burnet. The Viruses, G179.

Burstall. History of Mechanical Engineering, K54.

Burt. Field Guide to the Mammals, G236.

Burton. Systematic Dictionary of Mammals of the World, G237.

Buttress. World List of Abbreviations, A123.

Cahn. Introduction to Chemical Nomenclature, D154.

Cahn. Survey of Chemical Publication and Report to the Chemical
   Society, D3.

Cajori. A History of Mathematical Notations, B17.
Cajori. A History of Physics, C30.
California. University at Los Angeles. BMD: Biomedical Computer Programs, H68.
California Institute of Technology. Ranger V11 Photographs of the Moon, E67.
Callaham. Russian-English Chemical and Polytechnical Dictionary, D85.
Callataÿ. Atlas of the Moon, E68.
Callataÿ. Atlas of the Sky, E52.
Campbell. Statistics for Biologists, G60.
Carleton. Index to Common Names of Herbaceous Plants, G124.
Carpenter. Ecological Glossary, G149.
Carpovich. Russian-English Atomic Dictionary, C41.
Carpovich. Russian-English Biological & Medical Dictionary, G34.
Carrier. The Beginnings of Agriculture in America, J33.
Carter. Dictionary of Electronics, K94.
Carter. Dictionary of Inventions and Discoveries, K64.
CAST, A173.
Castiglioni. A History of Medicine, H43.
Castilla's Spanish and English Technical Dictionary, A76.
Catalog of the U.S. Geological Survey Library, F13.
Catalogue of Botanical Books in the Collection of Rachel McMasters Miller Hunt, G92.
Cecil-Loeb Textbook of Medicine, H115.
Celestial Handbook, E53.
Center for Research Libraries. Rarely Held Scientific Serials, A198.
Ceramic Abstracts, K23.
Chakravarti et al. Handbook of Methods of Applied Statistics, B83.
Challinor. Dictionary of Geology, F42.
Chalmers. Historic Researches, C31.
Chamberlin. Entomological Nomenclature & Literature, G156.
Chamber's Mineralogical Dictionary, F118.
Chambers' Technical Dictionary, A43.
Chase. Evolution of Modern Physics, C32.
Chase. Index to Grass Species, G125.
Chatten. Pharmaceutical Chemistry, H97.
Chem Sources, D163.
Chemical Abstracts, D14.
Chemical Abstracts. List of Periodicals with Key to Library Files, D15.
Chemical & Engineering News, D164.
Chemical-Biological Activities, D23.
Chemical Engineering Catalog, D165.
Chemical Engineering Practice, K145.
Chemical Engineers' Handbook, K144.
Chemical Guide to Europe, D166.

Chemical Guide to the U.S., D167.

Chemical Industry Directory and **Who's Who, D168.**

Chemical Markets Abstracts, D169.

Chemical Materials Catalog & **Directory of Producers,** D170.

Chemical Reviews, D44.

Chemical Society, London. Annual **Reports on the Progress of** Chemistry, F27.

Chemical Statistics Directory, **D171.**

Chemical Technology: an Encyclopedic **Treatment,** D63.

Chemical Titles, D16.

Chemical Who's Who, D177.

Chemisches Zentralblatt, D17.

Cherrington. Exploring the Moon Through Binoculars, E44.

Chilton's Auto Repair Manual, K174.

Chiu. Chinese-English, English-Chinese Astronomical Dictionary, E34.

Chorley et al. History of the Study of Landforms, F77.

Chow. Handbook of Applied Hydrology, F88.

Chu. Applied English-Chinese & Chinese-English Chemical Dictionary, D77.

Chymia, D50.

Clapper. Glossary of Genetics and other Biological Terms, G164.

Clark. Encyclopedia of Microscopy, A56.

Clark. Encyclopedia of Spectroscopy, A57.

Clark. Encyclopedia of X-rays and Gamma Rays, C54.

Clason. Elsevier's Dictionary of Metallurgy in Six Languages, K116.

Clavert. Encyclopedia of Patent Practice and Invention Management, A180.

Clifford. Electronics Data Handbook, K156.

Coats. Great Gardens of the Western World, G144.

Cochran & Haering. Solid State Physics, C54a.

Codlin. Cryogenics and Refrigeration, K24.

Collard. L'Astronomie et les Astronomes, E5.

Collective Numerical Patent Index to Chemical Abstracts, D22.

Collins. Complete Field Guide to American Wildlife, G209.

Comments on Astrophysics and Space Physics, E9.

Comments on Nuclear and Particle Physics, C24.

Comments on Solid State Physics, C25.

Commonwealth Agricultural Bureaux. Evaluation of the World Food Literature, J109.

Compton. Manual of Field Geology, F55.

Compton's Dictionary of the Natural Sciences, G28.

Compton's Illustrated Science Dictionary, A44.

Computer Abstracts, B60.

Computer & Control Abstracts, K25.

Computer Graphics, B69.

Computer Literature Bibliography, B61.

Computer Yearbook and Directory, B70.

Comrie. Civil Engineering Reference Book, K92.

Concise Encyclopaedia of Nuclear Energy, C49.

Condensed Chemical Dictionary, D64.

Condensed Computer Encyclopedia, B62a.

Condon & Odishaw. Handbook of Physics, C59.

Conference of Biological Editors. Style Manual for Biological
   Journals, G61.

Conn. Biological Stains, G51.

Considine & Ross. Handbook of Applied Instrumentation, K146.

Conway. The Weather Handbook, F107.

Corbin. An Index of State Geological Survey Publications
   Issued in Series, F14.

Corliss. Scientific Satellites, K132.

COSATI Report, A214.

Cotter. History of Nautical Astronomy, E14.

Cox & Pearson. Chemical Analysis of Foods, J116.

Craig. Bibliography of Encyclopedias and Dictionaries Dealing
   with Military, Naval & Maritime Affairs, K107.

Crane. Guide to the Literature of Chemistry, D4.

CRC Handbook of Biochemistry with Selected Data for Molecular
   Biology, D133.

CRC Handbook of Laboratory Safety, D134.

CRC Standard Mathematical Tables, B52.

Crew. The Rise of Modern Physics, C33.

Crowhurst. Electronics Reference Databook, K157.

Cumulative Index Medicus, H12.

Cumulative Index to Nursing Literature, H30.

Cunningham. Manual of Practical Anatomy, H85.

Current Agricultural Serials, J149.

Current Bibliography for Aquatic Sciences & Fisheries, F139.

Current Chemical Papers, D18.

Current Contents: Chemical Sciences, D19.

Current Contents: Life Sciences, G11.

Current Contents: Physical Sciences, A23.

Current List of Medical Literature, H9.

Current Literature in Traffic & Transportation, K26.

Current Medical References, H17.

Current Papers for the Professional Electrical and Electronics
   Engineer, K27.

Current Papers in Physics (CCP), C7.

Cusset. English-French and French-English Technical
   Dictionary, A61.

Czerni & Skrzýnska. English-Polish and Polish-English
   Technological Dictionary, K70.

Dairy Science Abstracts, J23.

Dalton. Sources of Engineering Information, K1.

Damm. Handbook of Biochemistry and Biophysics, D135.

Dana & Dana. System of Mineralogy, F124.

Dana & Hurlbut. Manual of Mineralogy, F123.

Daniels. Home Guide to Plumbing, Heating and Air Conditioning, J139.

Davenport & Rosenthal. Engineering, K47.

Davidsohn & Henry. Todd-Sanford Clinical Diagnosis by Laboratory Methods, H117.

Davies. French-English **Vocabulary in Geology and Physical** Geography, F49.

Davis. Forest Management, **J87.**

Davis et al. Microbiology, H92.

Davydov. Botanical Dictionary, G113.

Dawes. Hundred Years of **Biology, G22.**

Deep-sea Research and Oceanographic Abstracts, **F140.**

Deer. Rock-forming Minerals, F125.

Dental Abstracts, H24.

Denti. Dizionario Tecnico, A69.

De Sola. Abbreviations Dictionary, A58.

DeVries. French-English Science Dictionary, A62.

DeVries. German-English Medical Dictionary, H63.

DeVries. German-English Science Dictionary, A63.

DeVries & Clason. Dictionary of Pure and Applied Physics, C42.

DeVries & Herrman. English-German Technical and Engineering Dictionary, K67.

DeVries & Herrman. German-English Technical and Engineering Dictionary, K68.

Dictionary of Agricultural and Allied Terminology, J43.

Dictionary of Astronomical Terms, E23.

Dictionary of Chemical Engineering, K84a.

Dictionary of Chemistry and Chemical Technology in Six Languages, D89.

Dictionary of Geological Terms, F41.

Dictionary of Organic Compounds, D113.

Dictionary of Physics and Allied Sciences, C43.

Dictionary of Physics and Mathematics Abbreviations, B29.

Dictionary of Scientific Biography, A88.

Directory of Agricultural and Home Economic Leaders, J68.

Directory of Approved Internships & Residencies, H135.

Directory of British Scientists, A89.

Directory of Canadian Scientific and Technical Periodicals, A199.

Directory of Computerized Information in Science & Technology, A101.

Directory of Department of Defense Information Analysis Centers, A102.

Directory of Engineering Societies and Related Organizations, K209.
Directory of Federally Supported Information Analysis Centers, A103.
Directory of Medical Specialists Holding Certification by American Boards, H134.
Directory of Meteorite Collections and Meteorite Research, E75.
Directory of Palaeontologists of the World, F173.
Directory of Published Proceedings, A136.
Directory of Research Institutions & Laboratories in Japan, A113.
Directory of Selected Research Institutes in Eastern Europe, A114.
Directory of Special Libraries & Information Centers, A104.
Directory of Special Libraries in Israel, A105.
Dissertation Abstracts, A2.
Doane Agricultural Service. Farm Building Cost Handbook, J47.
Dodge et al. Diseases and Pests of Ornamental Plants, J73.
Doetsch. Microbiology; Historical Contributions, G177.
Dopkowski. Selected Bibliography of Indexing in Science and Technology, A140.
Dorian. Dictionary of Science & Technology, A64.
Dorian. Six-language Dictionary of Electronics, Automation and Scientific Instruments, K103.
Dorland's Illustrated Medical Dictionary, H53.
Dorland's Pocket Medical Dictionary, H53.
Downs & Jenkins. Bibliography: Current State and Future Trends, A3.
Dreisbach. Handbook of Poisoning, H98.
Dumbleton. Russian-English Biological Dictionary, G35.
Dunn. Short History of Genetics, G166.
Dunsheath. History of Electrical Engineering, K55.

Earth Science Reviews, F15.
East European Science Abstracts, D24.
Ebert. Physics Pocketbook, C60.
Ecology, G153.
Edwards. Bibliography of the History of Agriculture, J34.
Ehrlich & Murphy. The Art of Technical Writing, A147.
Elbert & Hyams. House Plants, J74.
Electrical & Electronics Abstracts (EEA), K27.
Electronics and Communications Abstracts, K28.
Ellis & Messina. Catalogue of Foraminifera, F158.
Elsevier's Dictionary of Aeronautics in Six Languages, K82.
Elsevier's Dictionary of Industrial Chemistry in Six Languages, D90.
Elsevier's Dictionary of Photography, K122.
Elsevier's Encyclopedia of Organic Chemistry, D112.
Elsevier's Lexicon of International and National Units, A78.
Elsevier's Medical Dictionary in Five Languages, H67.

Elsevier's Nautical Dictionary, K110.
Elsevier's Wood Dictionary in Seven Languages, J85.
Emin. Russian-English Physics Dictionary, C44.
Encyklopädie der mathematischen Wissenschaften mit Einschluss
 ihrer Anwendungen, B30.
Encyclopaedic Dictionary of Physics (Thewlis), C50.
Encyclopedia of Chemical Technology (Kirk-Othmer), D65.
Encyclopedia of Chemistry, D66.
Encyclopedia of Engineering Materials and Processes, K85.
Encyclopedia of Industrial Chemical Analysis, D95.
Encyclopedia of Plant Physiology, G103.
Encyclopedia of Polymer Science & Technology, D114.
Encyclopedia Science Supplement (Grolier), A52.
Encyclopedia of Textiles, J129.
Encyclopedia of Life Sciences, G40.
Encyclopedia of U.S. Government Benefits, A203.
Engineering Index, K13.
Engineering Plastics Monthly, K29.
Engineering Societies Library. Bibliography on Filing, Classi-
 fication and Indexing Systems, and Thesauri, K200.
Engineering Societies Library. Classed Subject Catalog, K14.
Engineers' Council for Professional Development. Selected
 Bibliography of Engineering Subjects, K15.
Engineers Joint Council. Thesaurus of Engineering and Scientific
 Terms (Project LEX), A204.
English-Hungarian Technical Dictionary, A67.
Ennion & Tinbergen. Tracks, G210.
Ensminger. Beef Cattle Science, J48.
Ergebnisse der Biologie, G15.
Ernst. Dictionary of Chemistry, D80.
Ernst & DeVries. Atlas of the Universe, E24.
Etherington. Nuclear Engineering Handbook, K194.
European Research Index, A115.
Excerpta Botanica, G83.
Excerpta Medica, H10.

Fairbridge. Encyclopedia of Atmospheric Sciences, E25.
Fairbridge. Encyclopedia of Geomorphology, F77a.
Fairbridge. Encyclopedia of Oceanography, F147.
Fairchild's Dictionary of Textiles, J130.
Farber. Great Chemists, D178.
Farber. Nobel Prize Winners in Chemistry, D179.
Farm Chemicals Hand Book, J49.
Farrall & Albrecht. Agricultural Engineering, K84.
Federation of American Societies for Experimental Biology.
 Federation Proceedings, G16.

Fenton & Fenton. The Fossil Book, F168.

FID News Bulletin, A225.

Field Crop Abstracts, J24.

Field. Foods in Health and Disease, J117.

Field. Patients are People, H75.

Finney. Introduction to Statistical Science in Agriculture, J57.

Fishbein. Modern Home Remedies and How to Use Them, H76.

Fisher & Peterson. The World of Birds, G222.

Fisher & Yates. Statistical Tables for Biological, Agricultural and
  Medical Research, G62.

Flammarion Book of Astronomy, E26.

Fletcher et al. Index of Mathematical Tables, B46.

Florkin & Mason. Comparative Biochemistry. D100.

Florkin & Scheer. Chemical Zoology, G196.

Florkin & Stotz. Comprehensive Biochemistry, D101.

Focal Encyclopedia of Photography, K120.

Fogel. Composite Index to Marine Science & Technology, F141.

Food Chemical News. Guide to the Current Status of Food
  Additives and Color Additives, J118.

Forbes. Forestry Handbook, J88.

Forestry Abstracts, J80.

Forsythe. Bibliography of Russian Mathematics Books, B7.

Fossilium Catalogus, F161.

Foster. Rock Gardening, J75.

Fouchier & Billet. Chemical Dictionary, D91.

Fowler. Guides to Scientific Periodicals, A192.

Fox. Physics and Chemistry of the Organic Solid State, D115.

Frandsen et al. Dairy Handbook and Dictionary, J50.

Franklin. Space Age Astronomy, E27.

Freedman. Handbook for the Technical & Scientific Secretary, A148.

Frey. Limnology in North America, F89.

Frisch. Nuclear Handbook, C62.

Fritzen. Rock-hunter's Field Manual, F126.

Fry & Mohrhardt. Guide to Information Sources in Space
  Science and Technology, K5

Fuel Abstracts and Current Titles, K30.

Gabrielson. The Fisherman's Encyclopedia, G231.

Gardner. Chemical Synonyms and Trade Names, D73.

Gardner. History of Biology, G23.

Garrison. An Introduction to the History of Medicine, H44.

Garrison & Morton's Medical Bibliography, H45.

Gas Abstracts, K31.

Gatto. Dizionario Technico Scientifico Illustrato, A70.

Gaylord & Gaylord. Structural Engineering Handbook, K152.

Geo Abstracts, F76.
Geochemical Prospecting Abstracts, F72.
Geology, 1888-1938, F31.
Geophysical Abstracts, F80.
Geophysical Directory, F82.
Georgano. Complete Encyclopedia of Motarcars, K104.
Geoscience Abstracts, F16.
Geoscience Information Society. Geologic Field Trip Guidebooks
  of North America, F1.
Gerth van Wijk. Dictionary of Plant Names, G114.
Gibbs. Identification Methods for Microbiologists, H93.
Gibbs & Shapton. Identification Methods for Microbiologists, G180.
Giese. Photophysiology, G47.
Gilliard. Living Birds of the World, G223.
Glasstone. Sourcebook of the Space Sciences, E28.
Glasstone. Sourcebook on Atomic Energy, C34.
Gleason. The New Britton & Brown Illustrated Flora of the
  Northeastern U.S., G126.
Gleason. Ultraviolet Guide to Minerals, F127.
Gleason et al. Clinical Toxicology of Commercial Products, H99.
Gleason & Cronquist. The Natural Geography of Plants, G151.
Glenn's Foreign Car Repair Manual, K175.
Glossary of Meteorology, F103.
Gmelin's Handbuch der anorganischen Chemie, D105.
Goldberg. Spanish-English Chemical & Medical Dictionary, D88.
Goldman. Guide to the Literature of Engineering, Mathematics,
  and the Physical Sciences, B1.
Good Housekeeping's Guide for the Young Homemakers, J140.
Goodman & Gilman. The Pharmacological Basis of Therapeutics,
  H100.
Gowen & Wheeler. Name Index of Organic Reactions, D123.
Graf. Exotica 3, G127.
Graf. Modern Dictionary of Electronics, K95.
Graham. The Basic Dictionary of Science, A45.
Grainger. Guide to the History of Bacteriology, G173.
Grant. Hackh's Chemical Dictionary, D67.
Grant. Handbook of Preventive Medicine & Public Health, H118.
Gras. History of Agriculture, J35.
Gray. Anatomy of the Human Body, H86.
Gray. Dictionary of Physics, C38.
Gray. Dictionary of the Biological Sciences, G29.
Gray. Encyclopedia of the Biological Sciences, G41.
Gray. Handbook of Basic Microtechnique, G52.
Gray. History of Agriculture in the Southern U.S., J36.
Gray. Manual of Botany, G128.

Great Britain. DSIR. Mathematical Tables, B53.
Great Britain. Meteorological Office. Meteorological Glossary, F104.
Green. History of Botany, 1860-1900, G102.
Greenwood & Hartley. Guide to Tables in Mathematical Statistics, B80.
Grignard et al. Traité de Chemie organique, D116.
Grimm. Recognizing Native Shrubs, G129.
Grollman. Pharmacology and Therapeutics, H101.
Gros. Constructional Engineering Dictionary, K93.
Grossman & Hamlet. Birds of Prey of the World, G224.
Guerra. American Medical Bibliography, H23.
Guide to Latin American Scientific & Technical Periodicals, A200.
Guidry & Frye. Graphic Communication in Science, A149.
Gunderson et al. Food Standards and Definitions in the U.S., J110.
Gunsalus & Stanier. The Bacteria, G181.
Gurr. Encyclopedia of Microscopic Stain, G53.
Guthrie & Miller. Home Book of Animal Care, J100.

Hackh's Chemical Dictionary, D67.
Haensch & Haberkamp. Dictionary of Agriculture, J44.
Hagerup & Petersson. A Botanical Atlas, G119.
Hall. Source Book in Animal Biology, G195.
Hall & Kelson. Mammals of North America, G238.
Hamby. The American Cotton Handbook, J131.
Hampel. Encyclopedia of Chemical Elements, D68.
Hampel. Encyclopedia of Electrochemistry, D128.
Hampel. Rare Metals Handbook, K189.
Handbook of Biochemistry & Biophysics, D135.
Handbook of Chemistry, D132.
Handbook of Chemistry & Physics, A79.
Handbook of Clinical Laboratory Data, H119.
Handbook of Electronic Engineering, K158.
Handbook of Food Additives, J119.
Handbook of Geochemistry, F73.
Handbook of Industrial Research Management, K147.
Handbook of Mathematical Tables, B54.
Handbook of Oceanographic Tables, F153.
Handbook of Physical Constants, F56.
Handbook of Thermophysical Properties of Solid Materials, D129.
Handbuch der Astrophysik, E39.
Handbuch der Pflanzenphysiologie, G103.
Handbuch der Physik, C51.
Handwörterbuch der Naturwissenschaften, A46.
Hanson. Dictionary of Ecology, G150.
Hanzak. The Pictorial Encyclopedia of Birds, G225.

Harmon. The Guide to Home Remodeling, J141.
Harned. Medical Terminology Made Easy, H54.
Harper Encyclopedia of Science, A47.
Harvard University. Arnold Arboretum. Catalogue of the Library, G93.
Harvard University. Gray Herbarium Index, G120.
Harvard University. Museum of Comparative Zoology. Library Catalogue, G190.
Haviland & House. Handbook of Satellites & Space Vehicles, K133.
Hawkins. Scientific, Medical & Technical Books Published in the U.S.A., A4.
Hawkins. Student's Engineering Manual, K123.
Haynes. American Chemical Industry, D51.
Haynes. Chemical Trade Names & Commercial Synonyms, D74.
Heald. The Making of Test Thesaurus of Scientific and Engineering Terms, A205.
Health Organizations of the U.S., Canada, & Internationally, H136.
Helminthological Abstracts, J94.
Henderson. Emergency Medical Guide, H77.
Henderson & Henderson. A Dictionary of Biological Terms, G30.
Henley's Twentieth Century Book of Formulas, D159.
Herald. Living Fishes of the World, G232.
Herbage Abstracts, J25.
Herland. Dictionary of Mathematical Sciences, B40.
Herner & Co. Recommended Design for the U.S. Medical Library and Information System, H69.
Herrick. Rocket Encyclopedia Illustrated, K74.
Herschman. Information Retrieval in Physics, C72.
Hertzberg. Repairing Small Electrical Appliances, J142.
Hewlett & Anderson. History of the U.S. Atomic Energy Commission, K56.
Hicks. Citizen's Band Radio Handbook, K165.
Highway Research Abstracts, K32.
Himmelsbach & Boyd. Guide to Scientific and Technical Journals in Translation, A155.
History of Science: An Annual Review, A35.
Hix & Alley. Physical Laws and Effects, C39.
Hoffleit. Catalogue of Bright Stars, E69.
Hoffmeister & Mohr. Fieldbook of Illinois Mammals, G239.
Hogben. Mathematics for the Million, B31.
Hogerton. The Atomic Energy Deskbook, C55.
Hohn. Dictionary of Electrotechnology, K100.
Hopkins. Standard American Encyclopedia of Formulas, D160.
Horn. Computer and Data Processing: Dictionary & Guide, B63.
Horn & Schenkling. Index Literaturae Entomologicae, G157.
Horner. Dictionary of Mechanical Engineering Terms, K105.
Horsfall & Diamond. Plant Pathology, G104.
Horticultural Abstracts, J26.

Hoseh & Hoseh. **Russian-English Dictionary of Chemistry and** Chemical Technology, **D86.**

Hospital Literature Index, **H32.**

Hospitals: Guide Issue, **H137.**

Houben-Weyl's Methoden der organischen Chemie, **D117.**

Hough. Scientific Terminology, A59.

Hough's Encyclopedia of American Woods, J82.

Houzeau & Lancaster. Bibliographie generale de l'Astronomie, E4.

Howard. Telescope Handbook and Star Atlas, E54.

Howell & Levorsen. Directory of Geological Material, F2.

Howell et al. Formula Index to NMR Literature Data, D25.

Huebner. Geology and Allied Sciences, F50.

Hungarian Central Statistical Office. Statistical Dictionary, B76.

Hungarian-English Technical Dictionary, A68.

Hunt & Groves. Glossary of Ocean Science and Undersea Technology Terms, F148.

Huntia, G90.

Hutchinson. The Genera of Flowering Plants, G130.

Hydata, F90.

Hyman. Astronautics Dictionary, E35.

Hyman. German-English, English-German Astronautics Dictionary, K80.

Hyman. German-English, English-German Electronics Dictionary, K101.

Hyman. The Invertebrates, G211.

IFI-Plenum Data Corporation. Directory, 1965-8, J120.

IFI-Plenum Data Corporation. Food and Color Additives Index, J121.

IGY World Data Center A: Oceanography. Catalogue of Data in World Data Center A, F154.

Illinois Agricultural Statistics. **Assessors'** Annual Farm Census, J59.

Illuminating Engineering Society. **IES Lighting Handbook,** K159.

Index Bergeyana, H95.

Index-catalogue of Medical & Veterinary Zoology, **G191.**

Index Chemicus, D26.

Index Kewensis Plantarum, G121.

Index Londinensis to Illustrations of Flowering Plants, Ferns and Fern Allies, G94.

Index Medicus, H12.

Index to Dental Literature, H25.

Index to Reviews, Symposia Volumes and Monographs in Organic Chemistry, D27.

Index to the Literature of American Economic Entomology, G158.

Index Translationum, A156.

Index Veterinarius, J95.

Industrial & Engineering Chemistry. Modern Chemical Processes, K86.

Industrial Research Laboratories of the U.S., A116.

Information on International Scientific Organizations, Services, & Programs for Chemists, Chemical Engineers and Physicists, D172.

Ingles. Mammals of the Pacific States, G240.

Institut Pasteur. Bulletin, G174.

Institute of Radio Engineers. IRE Dictionary of Electronic Terms and Symbols, K96.

International Abstracts of Biological Sciences, G12.

International Aerospace Abstracts, K33.

International Association of Agricultural Librarians & Documentalists. Current Agricultural Serials, J149.

International Association of Agricultural Librarians & Documentalists. Quarterly Bulletin, J7.

International Association of Volcanology. Catalogue of the Active Volcanoes of the World, F175.

International Astronomical Union. Les Observatoires astronomiques et les Astronomes, E76.

International Atomic Energy Agency. List of Bibliographies on Nuclear Energy, C12.

International Catalogue of Scientific Literature, A16.

International Chemistry Directory, D173.

International Code of Botanical Nomenclature, G117.

International Code of Nomenclature for Cultivated Plants, G118.

International Code of Zoological Nomenclature, G202.

International Conference on the Earth Sciences. Advances in Earth Science, F29.

International Congress Calendar, A128.

International Congresses and Conferences, 1840-1937, A137.

International Dictionary of Applied Mathematics, B32.

International Dictionary of Geophysics, F83.

International Dictionary of Physics & Electronics, C40.

International Directory of Back Issue Vendors, A106.

International Directory of Oceanographers, F155.

International Directory of Research and Development Scientists, A117.

International Encyclopedia of Physical Chemistry and Chemical Physics, C56.

International Federation for Documentation. Abstracting Services, A12.

International Federation for Documentation. Directories of Science Information Sources, A98.

International Geology Review, F28.

International Geophysical Year Special Committee. Annals of
  the International Geophysical Year, F43.
International Journal of Computer Mathematics, B62.
International Nursing Index, H31.
International Pharmaceutical Abstracts, H26.
International Plant Index, G122.
International Statistical Institute. Bibliography of Basic Texts
  and Monographs on Statistical Methods, B72.
International Tables for X-ray Crystallography, D136.
International Union of Biological Sciences. Index des Zoologistes,
  G215.
International Union of Geodesy & Geophysics. Abstracts, F84.
International Union of Geodesy & Geophysics. International
  Auroral Atlas, E70.
International Union of Pure & Applied Chemistry. Nomenclature
  of Inorganic Chemistry, D155.
International Union of Pure & Applied Chemistry. Nomenclature
  of Organic Chemistry, D156.
International Union of Pure & Applied Chemistry. Rules for
  IUPAC Notation for Organic Compounds, D157.
International Zoo Yearbook, G216.
Ireland. Index to Scientists of the World, A90.
Isis, A36.

Jablonski. Russian-English Medical Dictionary, H66.
Jackson. Glossary of Botanic Terms, G105.
Jackson. Guide to the Literature of Botany, G86.
Jacobs. Dictionary of Microbiology, G182.
Jacobs & Gerstein. Handbook of Microbiology, G183.
Jacobson. Encyclopedia of Chemical Reactions, D106.
Jaeger. Biologist's Handbook of Pronunciations, G31.
Jaeger. Source-book of Biological Names and Terms, G37.
Jahrbuch über die Fortscritte der Mathematik, B8.
James & James. Mathematics Dictionary, B33.
Jane's All the World's Aircraft, K181.
Jane's Fighting Ships, K182.
Jane's World Railways, K176.
Jansson et al. Handbook of Applied Mathematics, B55.
Jepson. Biological Drawings, G63.
Jerlov. Optical Oceanography, F149.
John Crerar Library. Catalog, A24.
John Crerar Library. List of Current Serials, A201.
Johnson. Satellite Environment Handbook, K134.

Johnson & Leone. **Statistics and Experimental Design in Engineering** and the Physical Sciences, **B84.**

Jolly. Official, **Standardized and Recommended Methods of** Analysis, **D96.**

Jonassen & Weissberger. **Technique of Inorganic Chemistry, D107.**

Jones. Flora of Illinois, **G131.**

Jones. Inventor's Patent Handbook, A181.

Jones. North American Radio-TV Station Guide, **K166.**

Jones & Schubert. Engineering Encyclopedia, **K65.**

Journal of Animal Ecology, G154.

Journal of Applied Chemistry, D20.

Journal of Chemical Documentation, D150.

Journal of the History of Biology, G26.

Journal of the Science of Food & Agriculture, J18.

Kaelbe. Handbook of X-rays, C67.

Kaestner. Invertebrate Zoology, G197.

Kaplan. Guide to Information Sources in Mining, Minerals, and Geoscience, F3.

Karpinski. Bibliography of Mathematical Works, B9.

Karpovich & Karpovich. Russian-English Chemical Dictionary, D87.

Karush. Crescent Dictionary of Mathematics, B34.

Kaye. Tables of Physical & Chemical Constants, A80.

Keenan & Atherton. The Journal Literature of Physics, C76.

Kéler. Entomologisches Worterbuch, G159.

Keller. Basic Tables in Chemistry, D137.

Kelley. Encyclopedia of Medical Sources, H128.

Kendall & Buckland. Dictionary of Statistical Terms, B77.

Kendall & Doig. Bibliography of Statistical Literature, B73.

Kent. Riegel's Industrial Chemistry, K87.

Kerchove. International Maritime Dictionary, K111.

Kerker & Murphy. Biological and Biomedical Resource Literature, G3.

Kett. Formation of the American Medical Profession, H46.

King. Dictionary of Genetics, G165.

King. History of the Telescope, E16.

King. Pictorial Guide to the Stars, E29.

King. Weeds of the World, G132.

Kingery. How-To-Do-It Books, K6.

Kingery et al. Men and Ideas in Engineering, K57.

Kingsbury. Poisonous Plants of the U.S. & Canada, G133.

Kingslake. Applied Optics and Optical Engineering, K121.

Kingzett's Chemical Encyclopedia, D69.

Kirby et al. Engineering in History, K48.

Kirchshofer, The World of Zoos, G217.

Kiss. Bibliography of Meteorological Satellites, F97.

Kit & Evered. Rocket Propellant Handbook, K135.

Klapper. Fabric Almanac, J132.

Kleczek. Astronomical Dictionary in Six Languages, E37.

Klempner. Diffusion of Abstracting & Indexing Services for Government-sponsored Research, A206.

Klerer & Korn. Digital Computer User's Handbook, B71.

Knight. Classical Scientific Papers - Chemistry, D52.

Kolthoff & Elving. Treatise on Analytical Chemistry, D97.

Konarski. Russian-English Dictionary of Modern Terms in Aeronautics and Rocketry, K81.

Kopal et al. Photographic Atlas of the Moon, E71.

Kopp. Geschichte de Chemie, D53.

Korn & Korn. Mathematical Handbook of Scientists and Engineers, B56.

Kotz. Russian-English Dictionary and Reader in the Cybernetical Sciences, B43.

Kotz & Hoeffding. Russian-English Dictionary of Statistical Terms, B78.

Kramer. Russian-English Dictionary of Astronomy, E36.

Krauch & Kunz. Organic Name Reactions, D124.

Kronick. History of Scientific and Technical Periodicals, A27.

Kuiper et al. Photographic Lunar Atlas, E72.

Kuiper & Middlehurst. The Solar System, E40.

Kuiper & Middlehurst. Stars and Stellar Systems, E41.

Kummel & Raup. Handbook of Paleontological Techniques, F169.

Kunz & Schintlmeister. Nuclear Tables, C63.

Kurtz & Edgerton. Statistical Dictionary of Terms & Symbols, B79.

Laboratory Handbook of Methods of Food Analysis, J122.

Laird. Engineering Secretary's Complete Handbook, A150.

Lamb & Lamb. Illustrated Reference on Cacti and other Succlents. G134.

Lancaster. Bibliography of Statistical Bibliographies, B74.

Landolt-Börnstein Zahlenwerte und Funktionen aus Naturwissenschaften und Technik, A82.

Landolt-Börnstein Zahlenwerte und Funktionen aus Physik, Chemie, Astronomie, Geophysik und Technik, A81.

Lange & Hora. Collins Guide to Mushrooms and Toadstools, G135.

Langman. Selected Guide to the Literature on the Flowering Plants of Mexico, G95.

Lanjouw & Stafleu. Index Herbariorum, G145.

La Rocque. Contributions to the History of Geology, F32.

Larousse Encyclopedia of Animal Life, G198.

Larousse Encyclopedia of Astronomy, E30.

Larousse Encyclopedia of the Earth, F44.
Laser Abstracts, K34.
Laser & Maser International, K35.
Laser Literature, K36.
Lauche. World Bibliography of Agricultural Bibliographies, J2.
Lawrence et al. Botanico-Periodicum-Huntianum, G79.
Lawrence. Literature of Taxonomic Botany, G78.
Lebedev & Fedrova. A Guide to Mathematical Tables, B47.
Lechevalier & Solotorovsky. Three Centuries of Microbiology, G178.
Leftwich. A Dictionary of Zoology, G199.
Lehrbuch der angewandten Geologie, F45.
Leibiger & Leibiger. German-English and English-German
  Dictionary for Scientists, A65.
Leicester. Source Book in Chemistry, 1900-1950, D55.
Leicester & Klickstein. Source Book in Chemistry, 1400-1900, D54.
Lejeune & Bunjes. Deutsch-Englisches, Englisches-Deutsches
  Worterbuch für Arzte, H64.
Lenhoff. Tools of Biology, G54.
Lepine. Dictionnaire Français-Anglais, Anglais-Français des
  Terms Médicaux et Biologique, H65.
Lettenmeyer. Dictionary of Atomic Terminology, C45.
Levitt & Marshall. Star Maps for Beginners, E55.
Lexicon of Geologic Names of the U.S. for 1936-60, F46.
Lichine et al. Alexis Lichine's Encyclopedia of Wines and Spirits,
  J123.
Liddicoat. Handbook of Gem Identification, F128.
Liebers. The Engineer's Handbook, K124.
Lincoln. American Cookery Books, 1742-1860, J111.
Linke. Solubilities, Inorganic and Metal-organic Compounds, D138.
Linton. Modern Textile Dictionary, J133.
List of Journals Commonly Cited in Geophysical Abstracts, F81.
Little. Checklist of Native and Naturalized Trees of U.S., J89.
Lockheed Aircraft Corp. Space Materials Handbook, K136.

McEntee. Model Aircraft Handbook, K137.
MacFall. Gem Hunter's Guide, F129.
McGill University. Dictionary Catalogue of the Blacker-Wood
  Library of Zoology and Ornithology, G192.
McGraw-Hill Basic Bibliography of Science and Technology, A48.
McGraw-Hill Encyclopedia of Science and Technology, A48.
McGraw-Hill Encyclopedia of Space, K75.
McGraw-Hill Modern Men of Science, A91.
McGraw-Hill Yearbook of Science and Technology, A50.
Machinery's Handbook, K177.
Macintyre & Witte. German-English Mathematical Vocabulary, B41.
Mackay. Bio-medical Telemetry, G64.

McLaughlin. Space Age Dictionary, K76.

McLean & Cook. Textbook of Theoretical Botany, G106.

MacRae's Blue Book, A163.

Maerz & Paul. A Dictionary of Color, C68.

Magie. Source Book in Physics, C35.

Maichel. Guide to Russian Reference Books, A5.

Maizell & Seigel. The Periodical Literature of Physics, C2.

Major. A History of Medicine, H47.

Malclès. Les Sources du Travail bibliographique, A6.

Malinowsky. Science and Engineering Reference Sources, A7.

Mallis. Handbook of Pest Control, J51.

Manning. Government in Science: the U. S. Geological Survey, F33.

Manufacturing Chemists' Assn. The Chemical Industry Fact
  Book, D174.

Marckworth. Dissertations in Physics, C11.

Margerie. Catalogue des Bibliographies geologique, F17.

Margerie. Critique et Géologie, F34.

Mark. Man-made Fibers, J134.

Markus. Electronics and Nucleonics Dictionary, K97.

Marler. Pharmacological & Chemical Synonyms, H107.

Marolli. Dizionario Technico, Inglese-Italiano, Italiano-Inglese, K71.

Marton. Foreign-language and English Dictionaries in the Physical
  Sciences and Engineering, A37.

Mason. Literature of Geology, F4.

Mason. Physical Acoustics, C57.

Massachusetts Horticultural Society. Library. Dictionary
  Catalog, G96.

Massachusetts Institute of Technology Library. Current Serials
  and Journals in the M. I. T. Libraries, K216.

Masters Theses in the Pure and Applied Sciences, A25.

Mathematical Biosciences, G65.

Mathematical Reviews, B10.

Mathematics of Computation, B49.

Mather & Mason. Source Book in Geology, F35.

Mather. Source Book in Geology, 1900-1950, F36.

Mathews. Catalogue of Published Bibliographies in Geology, F18.

Mayer. Chemical-technical Dictionary, D92.

Mead. Encyclopedia of Chemical Process Equipment, K88.

Medical Library Assn. Bulletin, H21.

Medical Library Assn. Directory, H138.

Medical Library Assn. Handbook of Medical Library Practice, H2.

Medical Progress, H36.

Medical Subject Headings, H12.

Meditsinskii Referativnyi Zhurnal, H18.

Meisel. Bibliography of American Natural History, F19.

Meites. Handbook of Analytical Chemistry, D139.

Mellon. Chemical Publications, D5.

Mellor. Comprehensive Treatise on Inorganic and Theoretical
  Chemistry, D108.

Menzel. Field Guide to the Stars and Planets, E56.
Menzel. Fundamental Formulas of Physics, C61.
Merck Index, D140.
Merck Manual of Diagnosis and Therapy, H102.
Merck Veterinary Manual, J101.
Merrill. The First One Hundred Years of American Geology, F37.
Merriman. Concise Encyclopedia of Metallurgy, K112.
Merritt. Standard Handbook for Civil Engineers, K153.
Metals Abstracts, K37.
Meteorological & Geoastrophysical Abstracts, F95.
Meteorological & Geoastrophysical Titles, F96.
Methods and References in Biochemistry and Biophysics, G76.
Methods in Medical Research, H37.
Methods of Biochemical Analysis, D102.
Methods of Experimental Physics, C52.
Mettler. History of Medicine, H48.
Miall. New Dictionary of Chemistry, D70.
Microbiology Abstracts, G175.
Midonick. The Treasury of Mathematics, B18.
Midwest Farm Handbook, J52.
Miller. Bibliography of the History of Medicine of the U.S. and
    Canada, H42.
Miller. Historical Introduction to Mathematical Literature, B19.
Miller. The Modern Medical Encyclopedia, H78.
Miller & West. Encyclopedia of Animal Care, J102.
Millington & Millington. Dictionary of Mathematics, B35.
Milne-Thomson. Russian-English Mathematical Dictionary, B44.
Mineralogical Abstracts, F115.
Minrath. Van Nostrand's Practical Formulary, D161.
Modern Chemical Engineering, K89.
Modern Chinese-English Technical and General Dictionary, K66.
Modern Drug Encyclopedia and Therapeutic Index, H108.
Mohlenbrock. The Illustrated Flora of Illinois, G136.
Moldenke & Moldenke. Plants of the Bible, G107.
Monthly Index of Russian Accessions, A157.
Montucla & Lalande. Histoire des Mathematiques, B20.
Moore. Agricultural Research Service, J38.
Moore. Book of Wild Pets, G212.
Moore. Handbook of Practical Amateur Astronomy, E45.
Moore. History of Chemistry, D56.
Moore. How to Clean Everything, J143.
Moore & Spencer. Electronics, K38.
Morgan. Airliners of the World, K138.
Morris. The Mammals, G242.
Morrison. Feeds and Feeding, J53.
Morton. How to Use a Medical Library, H3.

Moser. Space-age Acronyms, K77.
Mueller & Hartig. Geschichte und Literatur des Lichtwechsels, E17.
Multihauf. The Origins of Chemistry, D57.
Museums Directory of the U.S. and Canada, A118.

Naas & Schmid. Mathematisches Worterbuch mit Einbeziehung der theoretischen Physik, B36
Napier & Napier. Handbook of Living Primates, G241.
National Academy of Sciences. Biographical Memoirs, A92.
National Academy of Sciences. Catalogue of Data in World Data Center A, F57.
National Academy of Sciences. The Eastern European Academies of Sciences; a Directory, A124.
National Academy of Sciences. The Mathematical Sciences, B15.
National Academy Sciences-National Research Council. Recommended Dietary Allowances, J124.
National Agricultural Library Catalog, J11.
National Conference on Medical Nomenclature. Standard Nomenclature of Diseases and Operations, H56.
National Geographic Society. Wild Animals of North America, G213.
National Health Education Committee. Facts on the Major Killing and Crippling Diseases in the U.S. Today, H124.
National Research Council. Chemistry: Opportunities and Needs, D45.
National Research Council. Conference on Glossary of Terms in Nuclear Science and Technology, K117.
National Research Council. International Critical Tables, A83.
National Research Council. Oceanography Information Sources, F142.
National Research Council. Specifications and Criteria for Biochemical Compounds, D103.
National Research Council of Canada. National Technical Information Services, A107.
National Society of Professional Engineers. Directory of Engineers in Private Practice, K210.
Natural and Synthetic Fibers Abstract Service, J135.
Naturalists' Directory, G70.
Nature, A226.
Naylor & Naylor. Dictionary of Mechanical Engineering, K106.
Neave. Nomenclator Zoologicus, G204.
Nelson. Dictionary of Mining, K113.
Nelson & Nelson. Dictionary of Applied Geology, Mining, and Civil Engineering, F47.
Neville & Kennedy. Basic Statistical Methods for Engineers and Scientists, B85.
Neville et al. New German/English Dictionary for Chemists, D81.
New Illustrated Encyclopedia of Gardening, J76.

New York Academy of Medicine. Author and Subject Catalogs of the Library, H11.

New York Academy of Medicine Library. Catalog of Biographies, H129.

New York. Medical Library Center of N.Y. Union Catalog of Medical Periodicals, H143.

Newby. How to Find Out About Patents, A182.

Newman. The World of Mathematics, B37.

Nieuwenhuizen. Tropical Aquarium Fish, G233.

Nissen. Die botanische Buchillustration, G97.

Nobel Lectures, A93.

Nobel Lectures in Chemistry, D180.

Nobel Lectures: Physics, C36.

Nordenskiöld. The History of Biology, G24.

North Atlantic Treaty Organization. AGARD Aeronautical Multilingual Dictionary, K83.

North Carolina. University. An Introduction to the Literature of the Medical Sciences, H4.

Norton & Inglis. Star Atlas and Reference Handbook, E57.

Nuclear Data, C64.

Nuclear Science Abstracts, A167.

Nuclear Tables, C65.

Nuclear Theory Reference Book, C66.

Nutrition Abstracts & Reviews, J27.

Oakley. Frameworks for Dating Fossil Man, F170.

Oceanic Abstracts, F143.

Oceanic Index, F144.

Oceanography; the Weekly of the Ocean, F145.

Oliver. History of American Technology, K49.

O'Neill & O'Neill. The Unhandy Man's Guide to Home Repairs, J144.

Oppenheimer. Essays in the History of Embryology & Biology, G160.

Organic Reactions, D126.

Organic Synthesis, D127.

Oser. Hawk's Physiological Chemistry, H89.

An Overview of Worldwide Chemical Information Facilities and Resources, D151.

Owen. Handbook of Statistical Tables, B86.

Palmer. Fieldbook of Mammals, G243.

Palmer. Fieldbook of Natural History, A84.

Palomar Sky Atlas, E58.

Pandex, A21.

Pandex Current Index of Scientific & Technical Literature, A21.

Pannekoek. A History of Astronomy, E19.

Panshin et al. Textbook of Wood Technology, J83.
Parke. Guide to the Literature of Mathematics & Physics, B2.
Parker. Materials Data Book, K190.
Parker. Primer for Agricultural Libraries, J3.
Parr's Concise Medical Encyclopedia, H55.
Partington. A History of Chemistry, D58.
Pascal. Noveau Traite de Chemie minerale, D109.
Passwater. Guide to Fluorescence Literature, D28.
Patent Index to Chemical Abstracts, D21.
Patent Licensing Gazette, A183.
Patents for Chemical Inventions, D175.
Patterson. French-English Dictionary for Chemists, D79.
Patterson. German-English Dictionary for Chemists, D82.
Patterson et al. The Ring Index, D118.
Pearl. Gems, Minerals, Crystals and Ores, F119.
Pearl. Guide to the Geologic Literature, F5.
Pearson. Handbook of Lattice Spacings and Structures of Metals
   and Alloys, K191.
Pemberton. How to Find Out in Mathematics, B3.
Pennak. Collegiate Dictionary of Zoology, G200.
Pennsylvania. University. Catalog of the Edgar Fahs Smith
   Memorial Collection in the History of Chemistry, D59.
Perry. Engineering Manual, K125.
Pesticide Handbook - Entoma, J54.
Pesticides Documentation Bulletin, J19.
Peters. Check-list of Birds of the World, G226.
Peters. Dictionary of Herpetology, G246.
Peterson & McKenny. Field Guide to Wildflowers, G137.
Petit & Theorides. Histoire de la Zoologie, G194.
Petroleum Processing Handbook, K148.
Pharmacopoeia of the U.S.A., H109.
Philler et al. Annotated Bibliography of Technical Writing,
   Editing, Graphics, and Publishing, A141.
Physical Techniques in Biological Research, G48.
Physicians' Desk Reference to Pharmaceutical Specialties and
   Biologicals (PDR), H110.
Physics Abstracts, C8.
Physics and Chemistry of the Earth, F85.
Physics Express, C9.
Physikalische Berichte, C10.
Physiological Reviews, H38.
Pings. Plan for Indexing the Periodical Literature of Nursing, H144.
Planetary, Lunar and Solar Positions, E46.
Plant Breeding Abstracts, J28.
Plunkett. Handbook of Industrial Toxicology, H103.
Plutonium Handbook, K195.
Poggendorff's Biographisch-literarisches Handworterbuch, A94.
Polanyi. Technical Trade Dictionary of Textile Terms, J136.

Polymer Handbook, D141.

Polymer Science & Technology (POST), D29.

Popkin. The Environmental Science Service Administration, F99.

Porter. Bibliography of Statistical Cartography, F58.

Porter. Handbook of the Engineering Sciences, K126.

Porter & Spiller. The Barker Index of Crystals, D130.

Pough. Field Guide to Rocks and Minerals, F130.

Practical Handyman's Encyclopedia, J145.

Prager. Geschichte und Literatur des Lichtwechsels der veranderlichen Sterne, E18.

Principles of Zoological Micropalaeontology, F162.

Pritzel. Thesaurus Literaturae botanicae, G85.

Production Yearbook, J60.

Progress in Biophysics and Molecular Biology, G77.

Progress in Nuclear Energy, K46.

Progress in Oceanography, F146.

Progress in Physical Organic Chemistry, D46.

Progress in Physics, C26.

Progress in Solid State Chemistry, D47.

Pryor. Dictionary of Mineral Technology, K114.

Public Health Engineering Abstracts, K39.

Purdue University. Thermophysical Properties of High Temperature Solid Materials, D131.

Purdue University. Thermophysical Properties Research Literature Retrieval Guide, D30.

Pyle. General Class Amateur License Handbook, K167.

Quarterly Cumulative Index Medicus (QCIM), H12.

Quarterly Cumulative Index to Current Medical Literature, H12.

Quarterly Review of Biology, G17.

Radio Amateur's Handbook, K163.

Radio Handbook, K168.

Ransom. Range Guide to Mines and Minerals, F131.

Ransom. Fossils in America, F171.

Rasmussen. Readings in the History of American Agriculture, J39.

Rau. Dictionary of Nuclear Physics & Nuclear Chemistry, C46.

Ravin. The Evolution of Genetics, G167.

Reactor Handbook, K196.

Referativnyi Zhurnal, A19.

Regional Access to Scientific and Technical Information, A207.

Reinbek, Germany. Weltforstatlas, J90.

Reisigl. The World of Flowers, G146.

Remington. The Practice of Pharmacy, H111.

Research Centers Directory, A119.

Researches in Geochemistry, F74.
Reuss. Repertorium Commentationium, A14.
Review of Applied Entomology, G161.
Review of Applied Mycology, J29.
Review of Textile Progress, J137.
Reviews of Modern Physics, C27.
Rheology Abstracts, C13.
Rhodes et al. Fossils, F172.
Rice. Dictionary of Geological Terms, F48.
Richey et al. Agricultural Engineer's Handbook, K142.
Rickett. Wild Flowers of the U.S., G138.
Rider. Perpetual Trouble Shooter's Manual, K169.
Rider. Television Manual, K170.
Riley & Skirrow. Chemical Oceanography, F150.
RINGDOC. Derwent Pooled Pharmaceutical Literature Documentation:
  Abstracts Journal, H27.
Robb. Engineers' Dictionary, Spanish-English, English-Spanish, K72.
Robbins. Birds of North America, G227.
Roberts. Public Gardens and Arboretums of the U.S., G147.
Robinson. Hospital Administration, H120.
Rodd's Chemistry of Carbon Compounds, D119.
Roes & Kennedy. The Space-flight Encyclopedia, K78.
Rombauer & Becker. The Joy of Cooking, J125.
Romer et al. Bibliography of Fossil Vertebrates, F159.
Römpp. Chemie-Lexikon, D83.
Rosenau. Maxcy-Rosenau Preventive Medicine & Public Health, H121.
Rosin. Reagent Chemicals and Standards, D142.
Rossman & Schwartz. The Family Handbook of Home Nursing and
  Medical Care, H79.
Rothschild. A Classification of Living Animals, G205.
Roueche. Field Guide to Disease, H80.
Royal Geological & Mining Society of the Netherlands.
  Geological Nomenclature, F54.
Royal Society Mathematical Tables, B57.
Royal Society of London. Biographical Memoirs of Fellows, A95.
Royal Society of London. Catalogue of Scientific Papers, A15.
Rue. Sportsman's Guide to Game Animals, G214.
Ruffner & Thomas. Code Names Dictionary, A60.
Russian-Chinese-English Chemical & Technical Dictionary, D93.
Russian-English Dictionary of the Mathematical Sciences, B45.
Rutgers University. Bibliography of Research Relating to the
  Communication of Scientific and Technical Information, A142.

Sachs. History of Botany, G101.
SAE Handbook, K178.

Simonyi. Foundations of Electrical Engineering, K98.
Simpson. Geological Maps, F71.
Simpson. Principles of Animal Taxonomy, G206.
Singer. A History of Biology, G25.
Singer & Underwood. A History of Medicine, H50.
Singer et al. History of Technology, K50.
Singer. Short History of Scientific Ideas to 1900, A30.
Sinkankas. Van Nostrand's Standard Catalog of Gems, F132.
Sippl. Computer Dictionary, B64.
Sittig. Inorganic Chemical and Metallurgical Process Encyclopedia, K115.
Sitwell et al. Fine Bird Books, G218.
Sitwell & Blunt. Great Flower Books, G98.
Skinner. The Origin of Medical Terms, H57.
Sky & Telescope, E63.
Sky & Telescope. Telescopes, E31.
Smith. History of Mathematics, B23.
Smith. How to Find Out in Architecture & Building, K7.
Smith. Know-how Books, K8.
Smith. Mushroom Hunter's Field Guide, G140.
Smith. Rara Arithmetica, B11.
Smith. Small Arms of the World, K183.
Smith. Source Book of Mathematics, B24.
Smith. Sources for the History of the Science of Steel, K58.
Smith & Painter. Guide to the Literature of the Zoological Sciences, G185.
Smithells. Metals Reference Book, K192.
Smithsonian Institution. Astrophysical Observatory. Star Catalog, E60.
Smithsonian Institution. Smithsonian Meteorological Tables, F108.
Smithsonian Institution. Smithsonian Physical Tables, A85.
Snell & Dick. A Glossary of Mycology, G108.
Snell & Snell. Dictionary of Commercial Chemicals, D145.
Sneed et al. Comprehensive Inorganic Chemistry, D110.
Snyder. Complete Book for Gardners, J77.
Société Géologique de France. Bibliographie des Sciences géologique, F20.
Society of American Bacteriologists. Bergey's Manual of Determinative Bacteriology, H94.
Society of American Foresters. Forestry Terminology, J84.
Society of Experimental Biology and Medicine, G18.
Society of the Plastics Industry. Plastics Engineering Handbook, K149.
Sofiano. Russko-angliiskii Geologicheskii Slovar', F52.
Soils & Fertilizers, J30.
Solar Energy, K41.
Solar System Radio Astronomy, E32.
Solid State Abstracts, C14.

Solid State Physics, C28.

Souders. The Engineer's Companion, K127.

Source Books in the History of Science Series, A31.

Special Libraries Assn. Dictionary of Report Series Codes, A174.

Special Libraries Assn. Guide to Metallurgical Information, K9.

Special Libraries Assn. Guide to Special Issues and Indexes of Periodicals, A194.

Special Libraries Assn. Information Sources for the Biological Sciences and Allied Fields, G4.

Special Libraries Assn. Mutual Exchange in the Scientific Library and Technical Information Center Fields, A209.

Special Libraries Assn. Sources of Commodity Prices, A164.

Spector. Handbook of Toxicology, H104.

Spencer. Computer Programmer's Dictionary & Handbook, B65.

Stancy & Waxman. Computers in Biomedical Research, G67.

Stafleu. Taxonomic Literature, G80.

Standard Handbook for Electrical Engineers, K160.

Standard Handbook for Mechanical Engineers, K179.

Standard Methods of Chemical Analysis, D98.

Standard Nomenclature of Diseases and Operations, H56.

STAR, A168.

Statistical Theory and Methods Abstracts, B75.

Statistics Sources, A165.

Stearn. Botanical Latin. G116.

Stebbins. Field Guide to Western **Reptiles & Amphibians, G247.**

Stedman's Medical Dictionary, **H58.**

Steen. Dictionary of Abbreviations in Medicine, H59.

Stehli. The Microscope and How to **Use It, G56.**

Steinmetz. Codex Vegetabilis, G115.

Stephen & Stephen. **Solubilities of Inorganic and Organic Compounds,** D146.

Stephenson. The Gardener's Directory, **J69.**

Sterba. Freshwater **Fishes of the World, G234.**

Sterling & Pollack. Computers and the Life Sciences, G68.

Sternberg. How to Locate Technical Information, K2.

Stetka & Brandon. NFPA Handbook of the National Electrical Code, K161.

Steward. Plant Physiology, G109.

Strand. Basic Astronomical Data, E47.

Strand. An Illustrated Guide to Medical Terminology, H60.

Straub. A History of Civil Engineering, K59.

Strauss. Familiar Medical Quotations, H61.

Strock & Koral. Handbook of Airconditioning, Heating and Ventilating, K180.

Strong. Bibliography of Birds, G219.

Struglia. Standards and Specifications Information Sources, K206.

Struik. Concise History of Mathematics, B25.

Struik. Source Book in Mathematics, B26.
Sturtevant. A History of Genetics, G168.
Surrey. Name Reactions in Organic Chemistry, D125.
Survey of Biological Progress, G19.
Survey of Progress in Chemistry, D48.
Susskind. Encyclopedia of Electronics, K99.
Synthetic Organic Chemical Manufacturers Assn. SOCMA Handbook,
  D75.
Syracuse University Research Institute. Aerospace Structural
  Metals Handbook, K139.
System Development Corp. System Study of Abstracting and
  Indexing in the U.S., A210.
System Engineering Handbook, K150.
Szymanski. Infrared Band Handbook, D147.
Szymanski & Yelin. NMR Band Handbook, D148.

Taber. Cyclopedic Medical Dictionary, H62.
Tabulae Biologicae, G57.
"Take as Directed"; Our Modern Medicines, H82.
Taton. History of Science, A32.
Taylor. Guide to Garden Shrubs and Trees, J79.
Taylor. Physics, the Pioneer Science, C37.
Taylor & Taylor. Story of Agricultural Economics in the U.S., J40.
Taylor's Encyclopedia of Gardening, Horticulture, and Landscape
  Design, J78.
Technical Abstracts Bulletin (TAB), A169.
Technical Translations, A160.
Technology and Culture, K52.
Technology in Western Civilization, K51.
Telberg. Russian-English Dictionary of Geological Terms, F53.
Teilheimer. Synthetic Methods of Organic Chemistry, D120.
Thomas. Guide for Authors on Manuscripts, Proof, and Illustra-
  tions, A151.
Thomas' Register of American Manufacturers, A166.
Thomson. A New Dictionary of Birds, G228.
Thorndike. History of Magic and Experimental Science, A33.
Thornton. Medical Books, Libraries and Collectors, H51.
Thornton & Tully. Scientific Books, Libraries, and Collectors, A34.
Thornton et al. Select Bibliography of Medical Biography, H130.
Thorpe's Dictionary of Applied Chemistry, K90.
Today's Health, H83.
Todd-Sanford Clinical Diagnosis by Laboratory Methods, H117.
Topley & Wilson's Principles of Bacteriology & Immunity, G184.
Torrey Botanical Club. Index to American Botanical Literature,
  G87.
Trade Yearbook, J61.

Traité de Glaciologie, F91.

Traité de Paléobotanique, F165.

Traité de Paléontologie, F166.

Traité de Zoologie, G201.

Translations Register-Index, A158.

Translators & Translations, A159.

Treatise on Invertebrate Paleontology, F167.

Trollham & Whittmann. Dictionary of Data Processing, B66.

Turkevich & Turkevich. Prominent Scientists of Continental Europe, A96.

Turnbull. Scientific and Technical Dictionaries, A38.

Turner & Schmidt. Computers in Medicine: Bibliography, H70.

Turner. Technical Writer's & Editor's Stylebook, A152.

Ullmann's Encyklopädie der technischen Chemie, K91.

Ulrich's International Periodical Directory, A195.

UNESCO. Bibliography of Interlingual Scientific and Technical Dictionaries, A39.

UNESCO. Bibliography of Publications Designed to Raise the Standard of Scientific Literature, A144.

UNESCO. World Directory of National Science Policymaking Bodies, A120.

UNESCO. World Guide to Science Information and Documentation Services, A108.

Union Catalog of Medical Periodicals I, H143.

United Nations. International Bibliography on Atomic Energy, C15.

U.S. Air Force. Handbook of Geophysics, F86.

U.S. Aircraft, Missiles, and Spacecraft, K184.

U.S.A.E.C. Bibliographies of Interest to the Atomic Program, C16.

U.S.A.E.C. Subject Headings Used by the USAEC Division of Technical Information, A175.

U.S.A.E.C. Technical Books and Monographs, C3.

U.S.A.E.C. What's Available in the Atomic Energy Literature, C4.

U.S. Bureau of Customs. Merchant Vessels of the U.S., K185.

U.S. Bureau of Mines. Mineral Facts and Problems, F133.

U.S. Bureau of Mines. Minerals Yearbook, F134.

U.S. Bureau of the Census. 1964 U.S. Census of Agriculture, J62.

U.S. Coast & Geodetic Survey. Earthquake History of the U.S., F176.

U.S. Coast & Geodetic Survey. U.S. Earthquakes, F177.

U.S.D.A. Agricultural Statistics, J63.

U.S.D.A. Composition of Foods, J126.

U.S.D.A. Directory of Organizations and Field Activities, J70.

U.S.D.A. Foreign Agriculture, J64.

U.S.D.A. Guide to Understanding the U.S. Department of Agriculture, J41.

U.S.D.A. Handbook of Agricultural Charts, J67.

U.S.D.A. Index to Department Bulletins, Nos. 1-1500, J12.

U.S.D.A. Index to Farmers' Bulletins, Nos. 1-750, J13.

U.S.D.A. Index to Publications of the U.S. Department of
Agriculture, J15.

U.S.D.A. Index to Technical Bulletins, J14.

U.S.D.A. List of Available Publications of the U.S. Department
of Agriculture, J4.

U.S.D.A. Motion Pictures of the U.S. Department of Agriculture,
J5.

U.S.D.A. Preliminary List of References for the History of
Agricultural Science and Technology in the U.S., J42.

U.S.D.A. Silvics of Forest Trees of the U.S., K91.

U.S.D.A. Summary of Registered Agricultural Pesticide Chemical
Uses, J55a.

U.S.D.A. Workers in Subjects Pertaining to Agriculture in Land-
grant Colleges and Experiment Stations, J71.

U.S.D.A. World Agricultural Production and Trade, J66.

U.S.D.A. World Agricultural Situation, J65.

U.S.D.A. Yearbook of Agriculture, J55.

U.S.D.A. Library. Plant Science Catalog: Botany Subject Index,
G88.

U.S. Department of Commerce. Electric Current Abroad, K162.

U.S. Dispensatory and Physicians' Pharmacology, H113.

U.S. Division of Naval History. Dictionary of American Fighting
Ships, K108.

U.S. Environmental Data Service. Daily Weather Maps, Weekly
Series, F113.

U.S. Federal Council for Science and Technology. Committee on
Scientific & Technical Information (COSATI). Copyright Law
as It Relates to National Information Systems and National
Programs, A212.

U.S. Federal Council for Science and Technology. Committee on
Scientific & Technical Information (COSATI). Progress of
the U.S. Government in Scientific & Technical Communications,
A213.

U.S. Federal Council for Science and Technology. Committee on
Scientific & Technical Information (COSATI). Recommendations
for National Document Handling Systems in Science & Technology,
A214.

U.S. Federal Council for Science and Technology. Committee on
Scientific & Technical Information (COSATI). Standard for
Descriptive Cataloging of Government Scientific & Technical
Reports, A176.

U.S. General Services Administration. Federal Standardization,
K207.

U.S.G.S. Analytical Methods Used in Geochemical Exploration, F75.

U.S.G.S. Descriptive Catalog of Selected Aerial Photographs of Geologic Features of the U.S., F60.

U.S.G.S. Guide to Indexing Bibliographies and Abstract Journals of the U.S. Geological Survey, F21.

U.S.G.S. Index Map of the Subterrestrial Hemisphere of the Moon, F61.

U.S.G.S. Index to Geologic Mapping in the U.S., F62.

U.S.G.S. Index to Topographic Maps of the U.S., F63.

U.S.G.S. Publications of the Geological Survey, 1879-1961, F6.

U.S.G.S. Serial Publications Commonly Cited in Technical Bibliographies of the U.S. Geological Survey, F22.

U.S.G.S. Status of Aerial Photography in the U.S., F64.

U.S.G.S. Status of Geologic Mapping in the U.S., F65.

U.S.G.S. Status of Topographic Mapping in the U.S., F66.

U.S.G.S. Transcontinental Geophysical Maps Series, F67.

U.S. Government Organization Manual, A211.

U.S. Government Research & Development Reports, A170.

U.S. Government Research & Development Reports Index, A171.

U.S. Joint Chiefs of Staff. Dictionary of U.S. Military Terms, K109.

U.S. Library of Congress. Aeronautical and Space Serial Publications, K217.

U.S. Library of Congress. Bibliography of Snow, Ice and Permafrost with Abstracts, F92.

U.S. Library of Congress. Biological Sciences Serial Publications, G72.

U.S. Library of Congress. Directories in Science & Technology, A99.

U.S. Library of Congress. Guide to the World's Abstracting and Indexing Services in Science & Technology, A13.

U.S. Library of Congress. International Scientific Organizations, A109.

U.S. Library of Congress. MARC II Format, A216.

U.S. Library of Congress. Popular Names of U.S. Government Reports, A177.

U.S. Library of Congress. Project MARC, A215.

U.S. Library of Congress. Soviet Russian Scientific & Technical Terms, A73.

U.S. Library of Congress. U.S. IGY Bibliography, 1953-60, F23.

U.S. Military Academy, West Point. Atlas of Landforms, F78.

U.S. NASA. Astronautics & Aeronautics, K60.

U.S. NASA. Dictionary of Technical Terms for Aerospace Use, K79.

U.S. NASA. Earth Photographs from Gemini III, IV, V, F68.

U.S. NASA. Space Scientists & Engineers, K211.

U.S. NASA. Thesaurus, A178.

U.S. National Agricultural Library. Agricultural/Biological
Vocabulary, G32.

U.S. National Agricultural Library. Bibliography of Agriculture,
J8.

U.S. National Agricultural Library. Dictionary Catalog of the
Library, J10.

U.S. National Agricultural Library. Report of Task Force ABLE, J58.

U.S. National Agricultural Library. Serial Publications Indexed
in Bibliography of Agriculture, J9.

U.S. National Bureau of Standards. File Organization for a Large
Chemical Information System, D152.

U.S. National Bureau of Standards. The ISCC-NBS Method of
Designating Color Names, C69.

U.S. National Bureau of Standards. Measures for Progress, K61.

U.S. National Bureau of Standards. National Standard Reference
Data System, A217.

U.S. National Bureau of Standards. Publications, K208.

U.S. National Institutes of Health. Public Health Service Grants
and Awards, H125.

U.S. National Library of Medicine. Biomedical Serials, G73.

U.S. National Library of Medicine. Catalog, H14.

U.S. National Library of Medicine. Catalogue of Sixteenth
Century Printed Books, H16.

U.S. National Library of Medicine. Index Catalogue of the Library,
A17.

U.S. National Library of Medicine. List of Journals Indexed
in Index Medicus, H13.

U.S. National Library of Medicine. MEDLARS: 1963-67, H71.

U.S. National Library of Medicine. Monthly Bibliography of
Medical Reviews, H39.

U.S. National Library of Medicine. The National Library of
Medicine Classification, H72.

U.S. National Library of Medicine. National Library of Medicine
Current Catalog, H15.

U.S. National Library of Medicine. Russian Drug Index, H112.

U.S. Library of Medicine. Toxicity Bibliography, H29.

U.S. National Referral Center. Directory of Information Resources
in the U.S.: Federal Government, A110.

U.S. National Referral Center. Directory of Information Resources
in the U.S.: Physical Sciences, Biological Sciences, Engineering,
A111.

U.S. National Science Foundation. Current Research & Development
in Scientific Documentation, A218.

U.S. National Science Foundation. Federal Organization for
Scientific Activities, A219.

U.S. National Science Foundation. Nonconventional Scientific
& Technical Information Systems in Current Use, A220.

U.S. National Science Foundation. Providing U.S. Scientists
with Soviet Science Information, A161.

U.S. National Science Foundation. Scientific Information Activities
of Federal Agencies, A221.

U.S. Nautical Almanac Office. Air Almanac, E48.

U.S. Nautical Almanac Office. American Ephemeris and Nautical
Almanac, E49.

U.S. Naval Oceanographic Office. Glossary of Oceanographic
Terms, F152.

U.S. Office of Naval Research. Manual for Building a Technical
Thesaurus, A222.

U.S. Office of Water Resources Research. Water Resources
Research Catalog, F93.

U.S. Office of Water Resources Research. Water Resources
Thesaurus, F94.

U.S. Panel of the World Food Supply. The World Food Problem,
J127.

U.S. Patent Office. Development and Use of Patent Classification
Systems, A185.

U.S. Patent Office. General Information Concerning Patents, A184.

U.S. Patent Office. Index of Patents, A189.

U.S. Patent Office. Manual of Classification of Patents, A186.

U.S. Patent Office. Official Gazette, A187.

U.S. Patent Office. Official Gazette - Patent Abstract Section,
A188.

U.S. President's Science Advisory Committee. Science, Government
and Information, A223.

U.S. Public Health Service. Directory of Homemaker Services, J148.

U.S. Public Health Service. Directory of State and Territorial
Health Authorities, H139.

U.S. Public Health Service. Film Reference Guide for Medicine
and Sciences, H5.

U.S. Surgeon-General's Office. Library. Congresses, A130.

U.S. Weather Bureau. Climates of the States, F109.

U.S. Weather Bureau. Climatological Data for the U.S., F110.

U.S. Weather Bureau. Climatological Data; National Summary, F111.

U.S. Weather Bureau. Selective Guide to Published Climatic
Data Sources, F98.

U.S. Weather Bureau. World Weather Records, F112.

Uniterm Index to U.S. Chemical Patents, D176.

Universal Encyclopedia of Mathematics, B39.

Unlisted Drugs, H28.

Uphof. Dictionary of Economic Plants, G110.

Usher. Dictionary of Botany, Including Terms Used in Biochemistry,
Soil Science, and Statistics, G111.

Usovsky et al. Comprehensive Russian-English Agricultural
  Dictionary, J45.
UV Atlas of Organic Compounds, D121.
Uytenbogaardt. Tables for Microscopic Identification of Ore
  Minerals, F135.

Van Luik et al. Searching the Chemical and Chemical Engineering
  Literature, D6.
Van Nostrand's International Encyclopedia of Chemical Science, D71.
Van Nostrand's Scientific Encyclopedia, A49.
Van Nostrand's Standard Catalog of Gems, F132.
Vanderbilt. Amy Vanderbilt's New Complete Book of Etiquette, J146.
Vanders & Kerrs. Mineral Recognition, F136.
Vancouleurs. Astronomical Photography, E15.
Vaucouleurs & Vaucouleurs. Reference Catalogue of Bright
  Galaxies, E73.
Vehrenberg. Atlas of Deep Sky Splendors, E61.
Vehrenberg. Photographic Star Atlas, E62.
Velásquez & Nadurille. Obras de Consulta Agricolas en Español,
  J16.
Veterinarians' Blue Book, J103.
Veterinary Annual, J96.
Viewpoints in Biology, G20.
Viktorov et al. Short Guide to Geo-botanical Surveying, G141.
Vistas in Astronomy, E10.
Vital Notes on Medical Periodicals, H145.
Van Braun &Ordway. History of Rocketry & Space Travel, K62.

WADEX, K19.
Walden. Familiar Freshwater Fishes of America, G235.
Walford. Guide to Foreign Language Grammars and Dictionaries,
  A40.
Walford. Guide to Reference Material, A9.
Walker. Mammals of the World, G244.
Walker. Mathematics Essential for Elementary Statistics, B87.
Walther. Polytechnical Dictionary, K69.
Wang & Willis. Radiotracer Methodology in Biological Science, G58.
Ward. Geologic Reference Sources, F7.
Warren. Proposal: The National Library of Science System and
  Network, A224.
Watkins. Selected Bibliography of Maps in Libraries, F59.
Weatherwise, F114.
Webel. German-English Dictionary for Technical, Scientific,
  and General Terms, A66.
Webb. Bioastronautics Data Book, K140.

Weck. Dictionary of Forestry in Five Languages, J86.

Weed Abstracts, J31.

Weeks. Discovery of the Elements, D60.

Weigert & Zimmerman. Concise Encyclopedia of Astronomy, E33.

Weik. Dictionary of Computer Terms, B67.

Weinberg Report, A223.

Weinstein et al. Nuclear Engineering Fundamentals, K118.

Weisman. Technical Correspondence, A153.

Weissberger. Technique of Organic Chemistry, D122.

Wellcome Historical Medical Library, London. Catalogue of
 Printed Books in the Library, H19.

Weltforstatlas (World Atlas of Forestry), J90.

Westcott. Plant Disease Handbook, J56.

Westman et al. Reference Data for Radio Engineers, K171.

Wetmore. Song & Garden Birds of North America, G229.

Wherry. Automobiles of the World, K63.

White. Annotated Bibliography for History of Geology, F38.

White. Contributions to the History of Geology series, F39.

White. Reference Book of Chemistry, D72.

Whitford. Physics Literature, C5.

Whitman. First Aid for the Ailing House, J147.

Who's Who in Atoms, C74.

Who's Who of British Engineers, K213.

Who's Who in Engineering, K212.

Williams & Lansford. Encyclopedia of Biochemistry, D104.

Willis. Dictionary of the Flowering Plants and Ferns, G142.

Wilson. Comprehensive Analytical Chemistry, D99.

Winchell. Guide to Reference Books, A10.

Wit. Plants of the World, G123.

Witnah. History of the U.S. Weather Bureau, F100.

Wohlauer & Gholston. German Chemical Abbreviations, D84.

Woldman. Engineering Alloys, K193.

Wood. Introduction to the Literature of Vertebrate Zoology, G186.

Woods. The Naturalist's Lexicon, G38.

World Agricultural Economics & Rural Sociology Abstracts, J20.

World Aviation Directory, K214.

World Directory of Dental Schools, H140.

World Directory of Hydrobiological and Fisheries Institutions, G71.

World Directory of Marine Laboratories, F156.

World Directory of Mathematicians, B58.

World Directory of Medical Schools, H141.

World Directory of Veterinary Schools, J106.

World Health Organization. World Health Statistics Annual, H126.

World Index of Scientific Translations, A162.

World List of Future International Meetings, A131.

World List of Scientific Periodicals, A196.

World Medical Periodicals, H146.

World Meetings, Outside U.S.A. & Canada, A133.
World Meetings, U.S. & Canada, A132.
World Meteorological Organization. International Meteorological
  Vocabulary, F106.
World Nuclear Directory, C75.
World Space Directory, K215.
World Textile Abstracts, J138.
World Who's Who in Science, A97.
Wright & Wright. Handbook of Snakes of the U.S. & Canada, G248.
Wyszecki & Stiles. Color Science, C70.

Yakovlev. Handbook for Engineers, K128.
Yang et al. English-Chinese Dictionary of Chemistry and
  Chemical Engineering, D78.
Yates. How to Find Out in Physics, C6.
Yearbook of _____ series, H112.
Yearbook of Astronomy, E11.
Yearbook of International Congress Proceedings, A138.
Yearbook of Medicine, H40.
Yescombe. Sources of Information on the Rubber, Plastics
  and Allied Industries, K10.
Yugoslav Scientific Research Directory, A121.
Yukawa. Handbook of Organic Strucutral Analysis, D149.

Zalucki. Dictionary of Russian Technical & Scientific
  Abbreviations, A74.
Zeitschrift für Kristallographie, F116
Zentralblatt für Geologie und Palaontologie, F24.
Zentralblatt für Mathematik, B12.
Zentralblatt für Mineralogie, F117.
Zim & Smith. Reptiles & Amphibians, G249.
Zimmerman. Russian-English Translators Dictionary, A75.
Zimmerman & Lavine. Handbook of Material Trade Names, **D76.**
Zimmerman & Lavine. **Industrial Research Service's Conversion**
  Factors and Tables, **A86.**
Zinsser. Microbiology, **H96.**
Zoological Record, G193.
Zussman. Physical Methods in Determinative Mineralogy, F137.
Zwicky et al. Catalogue of Galaxies, E74.

www.ingramcontent.com/pod-product-compliance
Lightning Source LLC
Chambersburg PA
CBHW080529220326
41599CB00032B/6245